The Geometry of Universal Mind

by

Bob Mustin

Gridley Fires Books

Volume One

Other Books by Bob Mustin

A Reason to Tremble
A Place of Belonging
The Blue Bicycle
Sam's Place - Stories
We Are Strong But We Are Fragile
Collateral Damage and Stories
In This Love Together - A Memoir
The Third Reich's Last Eagle
The Perfection of Valor

Copyright © 2018 Bob Mustin.
All rights reserved.

ISBN: 978-1-64440-640-3

Library of Congress Control Number: 2018911700

All rights reserved. No part of this book may be produced or transmitted in any form, or by any means, electronic or mechanical, including photocopying, or by any information storage system, without permission in writing from the publisher or author.

First Published by *Gridley Fires Books*, 10/01/2018

Gridley Fires Books and its logo are trademarked
by
Gridley Fires Books Publishing

About the page format:

In keeping with the spirit of participation with those who might chance upon this book and make use of it, the author hopes those readers will become as interactive as possible with its ideas and text. The notations on the left of each page have been added to promote easy reference. The additional blank space on each, and the Notes pages at the book's end, can be used for reducing your ideas and thoughts to writing and for sketching. This is not intended to be a book read once, placed on a shelf, and forgotten. Instead, it's hoped readers will consider it a book of exploration into ideas – to extend the use and recognition of Geometry in modern life and in understanding of the created Universe.

TABLE OF CONTENTS

PREFACE..1

MIND AND CONCEPTUALITY

    1.01 Consciousness...4
    1.0 Mind...4
    1.1 Conceptuality..4
    1.2 The Alternating Expressions of Mind...8
    1.3 Exercises..14

THE HISTORY AND THE TRADITION

    2.0 The History of Geometry..15
    2.1 Geometric Analysis...19
    2.2 The Golden Mean...24
    2.2 A Word About Spirals...28
    2.4 The Mystery of Thoth...29
    2.5 Exercises..30

APPENDICES

CREDITS

INDEX

THE GEOMETRY OF UNIVERSAL MIND

The search for meaning

Preface:

All too often these days, I see unsatisfied people. They settle comfortably into work routines, into family life, but they always seem to be on the lookout for something else. Something more. As I've thought about that, I believe what's in short supply in all our lives has been **meaning**. Sure, a safe, secure position in a "whatever" career pays the bills, gives your family a comfortable place to call home, the opportunity to gain an education, to experience the wisdom of ages past. But many people have lost even these opportunities to live an examined life, and tragically, many of them turn bitter, even robotic. Still, some manage to see through repression, gain an understanding of their places in the world, and live lives that make sense to them – to their families, and to the communities surrounding them. This is rare, though, and that's unfortunate to the max.

This meaning thing; I realize it's hard to get a handle on it. Meaning in one's life doesn't just present itself gift-wrapped on the front porch. Great questions loom: How do we decide what in life is worth doing, and what is better left alone? How do we define ourselves, both as individuals and as part of a group of like-minded – or not-so-like-minded people? Once we become comfortable in what and who we are, what's our role in the world supposed to be?

The achievements of the past century prove beyond doubt that there's nothing we cannot achieve, given the will to do it and the resources to make it happen. But we now live in a world in which we're constantly confronted with choices. Nothing, it seems, is out of bounds; there's no clear path to happiness and inner satisfaction, so it's not surprising that we see so much compulsion, so much addiction, in our families, our colleagues, our neighbors, as we struggle to make something – anything – work in giving our lives a foundation of meaning.

Well. How do we tiptoe through this social minefield, each of us groping as if in a darkened room for substance? Let me suggest that we first learn the art of not-doing. Find time each day to rest, to contemplate, not just the day's events, but life itself. For just a few minutes, free yourself from your responsibilities, even your thoughts. Spend a little time sitting under a tree enjoying nature. Listen to the bird songs, feel the wind as it streams past you. Or learn breathing exercises or practice meditation.

It has been just such contemplation that led me to write this

THE GEOMETRY OF UNIVERSAL MIND

Mind, consciousness, and Geometry

book, to speak about personal direction, the urge toward meaning. About focusing on the things in our lives that seem significant, the pathways to something meaningful. There's much work to be done in making one's life seem real, though. We need to at least provisionally answer the big questions mentioned above. We need to know what makes us tick. We need glimpses of the Big Picture, i.e., how we see ourselves fitting into the grander scheme of things. And so, put simply, meaning is this: knowing oneself and how you fit into the organism of the rest of the world. Is achieving this a doable proposition in such a confused and muddled world? I'm sure it is or I wouldn't be writing this. But in order to do so we must begin by reaching deep inside to find the core of ourselves.

What tools do we have at our disposal in coming to this discovery of self? There are many. Psychological counseling. Religious practice. Meditation. Exercise and health awareness. General education. Social interaction. And many more. So why have I chosen to promote Geometry as an avenue to meaning?

It's becoming clearer to me each day that Geometry permeates everything we do, everything we are, everything in life that we put ourselves into. Even our physical attraction to others is underlain in large part by the symmetry and proportioning of the human body - and that's Geometry. Our minds constantly seek and find patterns in nature, in the behavior of others, and these patterns, while almost always imperfect and a little off-center, are Geometry. The structure of plant life, the socialization of ants and bears and coyotes and dolphins, the shapes of mountain ranges and pyramids, the constantly changing weather patterns – all these are built on Geometry. In fact, over the centuries we've seen many such patterns repeating in life and recognized the Geometry lurking in them, something I see referred to as Sacred Geometry. (A wry note: While I too see these patterns and the Geometry that supports them, I edge away from using the term sacred. With tongue in cheek, I ask: Have you ever heard of Profane Geometry? Why not, then, leave the subject out of the mystical and mysterious and refer to it simply as Geometry?)

All right, you say, this Geometry of yours is just a way of seeing the world around us, and to that I reply, "Yes! It's the way our minds work!" Once we realize that, we begin to notice an intelligence at work in the patternings both within and outside us. And so I propose that the consciousness, the intelligence at work within us is also at

THE GEOMETRY OF UNIVERSAL MIND

Language and meaning

work in everything we examine, no matter how remote or staid the things we perceive are. Realizing that, I think, will eventually involve a rediscovery of the grandest scope of mind, which I call **Universal Mind**, something I might refer to as the active verb of consciousness. Universal Mind is the playground of creation; it's where consciousness finds expression.

To back up a step, if the imprint Universal Mind makes on consciousness' blank slate is composed of Geometric shapes, how does this imprint apply to me, to meaning? It will be the task of this book series to answer that question. We'll begin by reminding that philosophers employed Geometry for millennia as symbols to help us understand the makeup of nature, of form's enduring realities. In other words, symbols that took on meaning. Meaning for our individual lives and meaning about us in the world we inhabit. But there's also a problem: too often these Geometric symbols, the sense of meaning they're intended to convey, were cloaked in ancient, esoteric language and ideas. As our many scientific fields developed and were explored, the wisdom and understanding these disciplines held was quietly separated by the different languages of physics, thermodynamics, structural design, fluid mechanics, surveying, and on and on. And that makes using the tools of science in the quest for meaning an extra challenge. So we'll turn once more to Geometry, the old philosopher's stone.

As a working engineer, and no doubt because of the language and mathematical variations from scientific discipline to scientific discipline, I was slow to understand that Geometry plays a part in underpinning all the fields science chose to explore. My thanks to Buckminster Fuller, P. D. Ouspensky, David Bohm, and others whose work inspired me and brought me to a fuller (pun intended) extension of Geometry's place, not only in technology, but in philosophy, in spirituality, and quite possibly in the broader spectrum of the arts.

My ideas here may suffer from scientific inexactness, but don't let that throw you. I'm a conceptualist by nature, and the general always suffers from lack of the specific. Using this book as a workbook may help fill in many specific blanks in your thinking. And perhaps these efforts, as we rediscover meaning in our lives and in the world about us, will help us rise to the next level of human awareness.

~ Bob Mustin ~

Mind
and
Conceptuality

THE GEOMETRY OF UNIVERSAL MIND

1.01 Consciousness – It is awareness. The creative plasma of Universe. It is life itself.

1.02 Mind – It's the patterning embedded into consciousness. It's consciousness' partner in the process of creation.

1.1 Conceptuality

Today we think of Geometry simply as a very visual form of mathematics, but its historical role has been so much more than that. Still, math is a place to begin. If we're to use Geometry to help understand ourselves and our world, we should first agree roughly on how we're to think of its shapes. As it turns out, there are two solid strategies available:

Generalized – In this type of modeling, we give broad meaning to Geometric symbols so we can be as all-inclusive as possible, i.e., so we can have them apply as much as possible to the entirety of the Universe, to the entirety of thought. The strength in generalization? Simplicity. Generalized models are profoundly lacking in detail. They do, however, tend to aid in the synthesis or combining of ideas. This Geometry tends to be holistic, far-reaching. In this strategy, a certain shape attracts raw, chaotic data and finds ways to have it apply to the concept behind that shape.

Specific – Still, special case situations will come up that won't easily fit these generalized models. Specific modeling tends to be based on evidence or data we gather, i.e., we construct a Geometric shape based on the data. This makes these shapes rather busy or complex, and they're highly dependent on special conditions and relationships. Their strength lies in accurately representing detail. There's a lot of beauty in this type of modeling, as we'll see when we get to Chaos Theory in a subsequent volume, but to a degree it can sacrifice a clear vision of the overall concept the Geometric shape has at its basis. Too, these models can be dynamic, i.e., as one part of the model changes, other parts may change.

The use of Geometry in this book will tend toward the generalized model. Why? Because it's simple to represent and simple to understand – a good place to start. Once you understand the generalized version, it'll be easier to understand the more specific types as you add complexity. And allow them to be dynamic.

Modeling strategies

Generalized

Specific

The generalized model is chosen

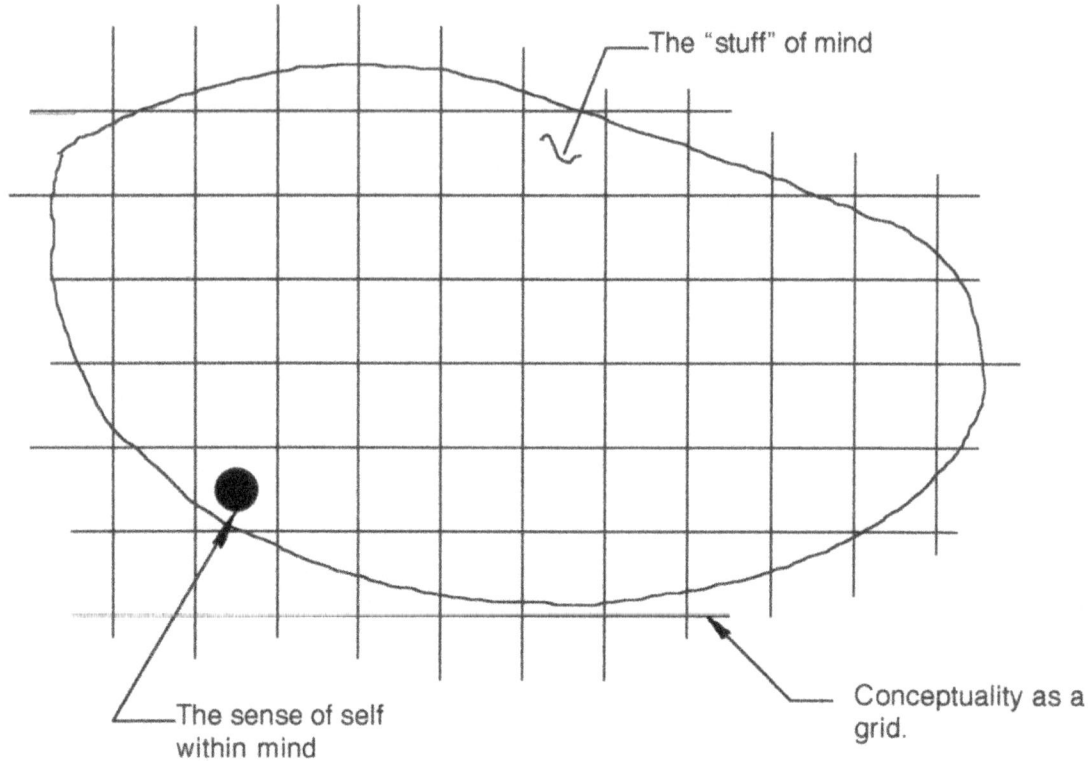

- The "stuff" of mind
- The sense of self within mind
- Conceptuality as a grid.

Upon the "stuff" of mind a grid of conceptuality is superimposed, which is meta-physical. This grid allows mind to experience itself as diversity subject to free will. It is a 4D holographic experience, much like the 2D experience of reflections in a mirror. The grid is an attractor, always connecting diversity to its origins in the crucible of mind.

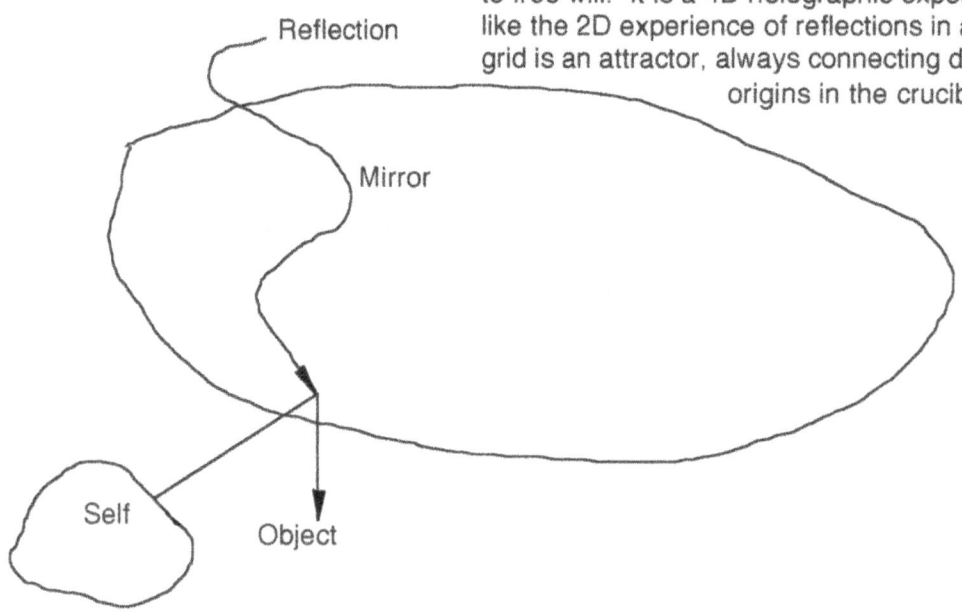

Figure 1

THE GEOMETRY OF UNIVERSAL MIND

Nature orients itself about a grid of patterns, or attractors.
See Figure 1

Speaking of things dynamic...

You'll understand, even from the History and Tradition section of this volume, that in viewing the natural world's makeup, its phenomena are never the same; they're always changing, if only slightly. Still, patterns or shapes become visible, and once they're established, the data points seem to cling to or cluster about them. These phenomena are generated in association with enduring but fluctuating patterns. But these patterns don't exist in the world of the senses. They do give a sense of order to these phenomena, though, no matter how chaotically and out of sync the phenomena seem to appear. These patterns are conceptual structures within consciousness. Here we call them Geometry. As we'll see in a subsequent volume, Chaos Theory has a different name for them: **attractors**.

We can prove, from examinations of numerous instances in nature, that these patterns, or attractors, don't exist on the level of static, natural phenomena. But they do affect the behavior of natural phenomena from the time they come into existence until these phenomena wane and disappear.

How does the randomness, the freedom, of the Universe's natural phenomena manage to keep its willy-nilly nature and still drift slowly toward coherent behavior? Toward beauty? Toward meaning? Toward the great abstraction called love? Clearly an agreement of sorts has been struck that allows the abject abandon, the complete expressions of freedom, to coexist with the restrictions of patterned structure and behavior.

Chaos coexists with order... the physical with the meta-physical.

If the Universe's phenomena and the chaos they initially exude as they come into physical existence are one side of their nature, then the still, silent patterning attracting these phenomena are a complementary, or metaphysical flip-side. Meta, from the Greek, means "along with." Meta-physical, then, is the other "half" of the physical, i.e., the part without which it would not be whole.

The physical consists of tangible phenomena, which we can only describe with a satisfactory degree of accuracy through their relationships with one another. The meta-physical is the symbolic nature of those relationships. The meta-physical is a sort of gravity, a watershed point about which the physical spins and orbits and pulses. The meta-physical is at the juncture of the physical – the transitory, here-today, gone-tomorrow – and the transcendental, the eternal. The meta-physical is a function of pure conceptual space, i.e., it isn't defined in any way by time, except in its symbolic forms

Meta-physical is conceptual...time is reduced to symbolism.

THE GEOMETRY OF UNIVERSAL MIND

(more on this later). If time were to be voided completely, however, there would be nothing but the potential of eternity – a great gulp of infinite possibilities. But as time begins to leave the phenomenal world, slowly revealing the symbolic qualities of the meta-physical, it's possible to devise the blueprints of the physical creation. The place where this happens is the world of the conceptual. And we see this world as Geometric shapes.

The pulsing, spinning, dervish-like world of the physical always moves toward the symmetry found in conceptuality but never rests there. The eternally-in-motion nature of the physical is always asymmetric, lopsided. Were it to attain the symmetry of its partner in creation, it would lose its dynamic nature and would have to be recreated. If that's confusing, think of a Sufi dancer. No? Say a hula dancer, then! Both dancers are constantly in motion, but both cling to the human shape they walked into the house with. If either stopped dancing, moving, breathing, etc., they would be seen as dead – or more palatably, as inert as a statue. **The physical/phenomenal can only coexist with the transcendental/eternal with the meta-physical as a transition form between them.**

It's important to understand that what science is currently discovering at the root of its ongoing examination of the natural world is the conceptual. And that's where Geometry's beginnings are! In this sense it's the blueprint, the basis of creation.

But please understand as well that while conceptuality sets the stage for phenomenal life, the erupting-then-shrinking, chaotic creation of the physical drags images of the conceptual into the physical's realm. One pushes and pulls the other into a dynamic balance. The cosmic joke of this arrangement is that while fundamental evolution in both is possible, such changes initiate more easily from the physical. I.e., the physical's changeable, energetic nature finds it easier to tweak the conceptual than the opposite. Considering both scenarios, each would look like this:

Physical changes the conceptual: Even though the conceptual initially shows itself in perfect symmetry, repeating its patterns infinitely, it is flexible and forgiving. Because of that, the energy that flows through the physical realm can set up standing waves that, if they're held long enough or powerfully enough, they can change the conceptual pattern. Picture it this way: you represent the conceptual by a series of identical balloons that

---

There are only two ways to change things:

Change begins in the physical realm.

THE GEOMETRY OF UNIVERSAL MIND

are inflated identically and arranged in rows and columns. You then stick a pin in some, deflate some slightly, and overinflate some. Doing this mimics the willy-nilly nature of energy coursing through the physical realm. These actions have then changed the shape of the conceptual pattern Geometry but not its integrity. Energy always seeks the most efficient pathway of getting from point A to point B, though. So once the conceptual has been changed, and energy flows even more erratically, the conceptual changes again. The conceptual can change continually, then, the conceptual and the physical anchoring each to the other.

Conceptual changes the physical: Ever heard of an account in which the heavens parted? This scenario would be a lot like that. Here, some master force (which would be tantamount to the godly) would change the Geometric patterning. Still, this patterning would have to be some expansion or collapsing of the original patterning, because it would still have to be the most efficient one and the one that's the most flexible and forgiving, even though it's able to withstand the most powerful energies from the physical.

> Change begins in the conceptual realm.

Clearly, then, the latter isn't likely to happen very often. This would indicate that the very nature of the Universe would have to change for this to happen (once again – not to say that it couldn't). Perhaps the conscious awareness of humanity as a whole, in league with even higher forces, could make such a change happen through a focused use of passion, will, and love.

And while we're at it, **if we were to define love in the language of science**, we would have to say it's an expression of gravity, or the pull toward maximum efficiency, ultimately toward forever, the eternal. If we set the conceptual as the standard for maximum efficiency, we must realize that efficiency – the process of how maximum strength **and** flexibility are reached – is always a function of what we collectively believe is possible. If what is possible has passion, will, and love as its basis,* then we can confirm that the physical and the meta-physical must move toward the truly eternal. Passion, will and love, then, organize the physical (and okay, they can conceivably reorder the heavens). If the outer expressions of passion and will must always be most effective in such a project, then "most

> Passion, will, and love
>
> Love moves creation toward efficiency, toward eternity... and eternity toward creation?

---

* For the sake of this usage, passion is unfocused emotion, will is the human faculty used in initiating action, and love is the internal feeling of Oneness with all.

effective" equates to love, the inner and all-embracing force. Thus the Universe's movements can change from outside in and inside out. And always by way of love.

1.2 The Alternating Expressions of Mind

See Appendix "A"

Mind, with the Universal range of its complements, is not a static field of expression. It allows for a multitude of freedoms, which can be empirically catalogued from human experience. We continue to discover every day that our thoughts are always on display through the senses. "As above, so below," and "the macrocosm is the microcosm" are tenets we can take very literally. Even as we adapt to the cycles of the seasons, we see the tides ebb and flow, we sleep and waken with the cycles of day and night created by the spinning of the planet on its axis.

The greater sense of Mind creates cyclic activity in order to keep its creations "lively," i.e., to keep them from decaying quickly into inert objects and phenomena and "vacuumed" up into Mind's unconscious, where they're dumped into other regions of the Multiverse (a grander sense of Universe vibrating at frequencies other than ours) and thus given new life. As we'll discover, no phenomenon can be separated from the matrix or field within which it occurs. To offer a less abstract example, no piece of magnetized steel could be separated from the magnetic field of which it's a component.

Complementary cognitive processes: intuition & rationality.

In that sense, we know we all possess complementary cognitive processes: the rational, left brain process of intellect, and the holistic, non-linear, right brain process of intuition. If these two processes working within each of us are the microcosm, how do we envision its macrocosm?

A skilled astronomer, or even a capable astrologer, will tell us about the precessional effect of the Earth's axis, tilted as it is twenty-three degrees away from the perpendicular to its plane of orbit around the sun. The ancient cultures of Earth, tapped into seasonal cycles, equinoxes and solstices, noticed that over time there were unique flavors to these cycles. These ancient investigators noticed that the sun, moon, and planets appeared in different constellations as Earth's seasons cycled by. Over eons they noticed that these heavenly bodies slowly shifted from a constant position relative to the constellations in each season. For instance, if a given planet appeared in Pisces in a given season, they noticed that as the years rol

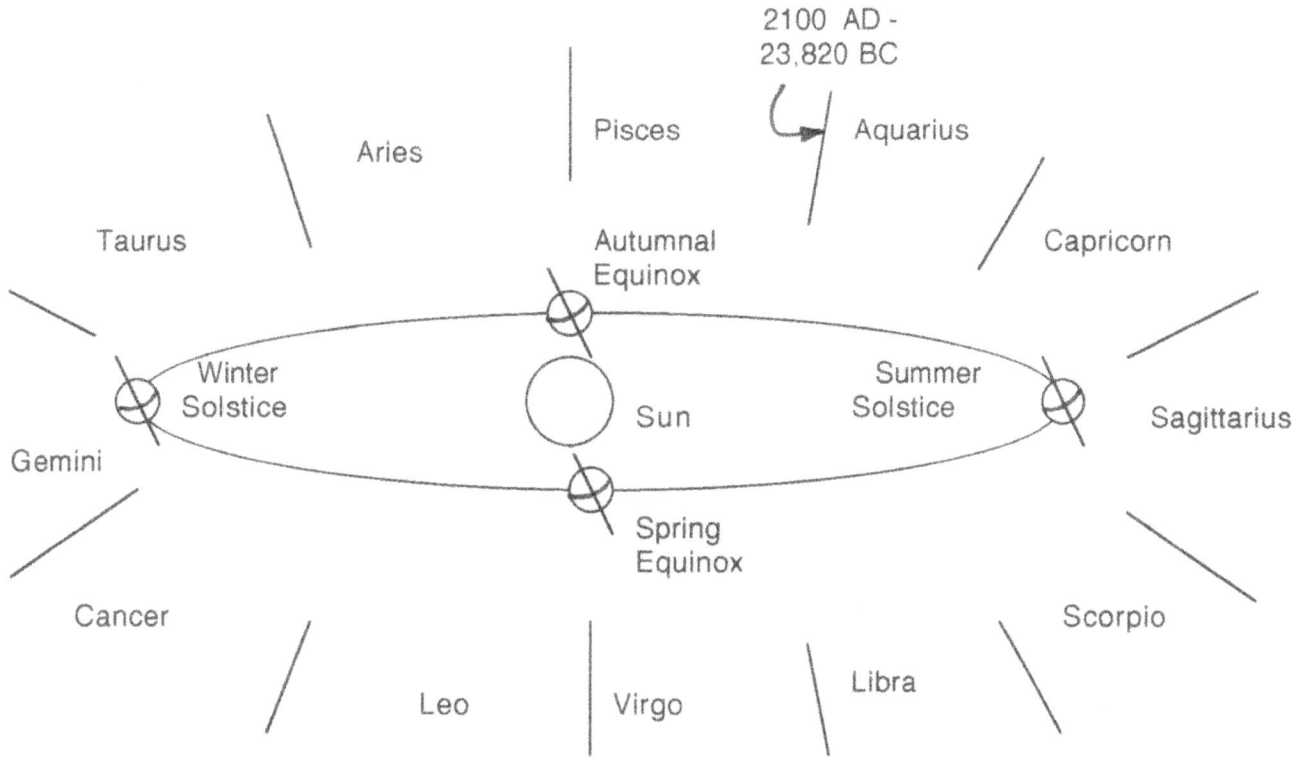

Each of the twelve subdivisions of the precessional orbit of Earth about the Sun is 2160 years. The total time of the orbit is: 12 x 2160 years, or 25,920 years.

Figure 2

THE GEOMETRY OF UNIVERSAL MIND

Precession.

Changes ...Mind's organization every 2160 +/- years...

...and one full cycle completes over 25,920 years.

Alterations in Earth cultures

See "Appendix "B"

Left brain/right brain.

Greek culture

Thales

-led by, the position of the planet at this time slowly shifted. After many years it would move completely away from Pisces and into the constellation of Aquarius.

This "drag" in the cycles is due to the precessional effect caused by the Earth's wobbling spin axis as it orbits the sun. It's acknowledged by astrologers in divining the various characteristics of the astrological signs with respect to time. It has also been measured by astronomers as being fifty-two seconds of arc in one year, or about one full degree in seventy-two years. Therefore, over 2160 years the location of the planet moves by one full zodiacal house, or one of twelve, more or less equal segments of the orbit of Earth around the sun. Tracking this eccentricity of Earth's polar spin, we may notice that the position of the planet returns to its original stellar placement (in Pisces, for example) once every 25,920 years.*

But we were talking about the alternating expressions of Mind. The significance of this stellar geography lesson is that we also notice changes in Earth's cultures, corresponding roughly to the 2160-year periods. I believe these changes are caused by the precessional effect of the Earth as it moves through thirty degrees of its own orbital arc or the astrologer's one zodiacal house. These cultural shifts seem to be alternating ones, too, back and forth between socialized expressions of left brain, or rational mind, and right brain, or intuitive mind. Admittedly there is a limited possibility of accurately researching this supposition, given the dearth of thorough historical records of ancient cultures. Research in this field, which I believe would support such a supposition, would be most edifying to future generations in understanding the changes occurring within the Earthbound human psychology as it modifies itself every two millennia or so.

Certainly the history of the past three thousand years supports this idea. No such study can avoid the Greek culture, which rose to prominence some 3000 to 2500 years ago. The life of Greek citizen Thales (circa 580 BCE) exemplified the transitional social changes occurring during that era. In early Greek Europe, tribal societies predominated. Their arts and religions were dominated by mythological thinking and concepts, i.e., making abstract ideas seem dressed in human clothes and habits. Which is to say that where today's scientists talk of quantum mechanics and the relationship between virtual particles, the seers of that time sensed the same ab

---

* See http://astro.wsu.edu/worthey/astro/html/lec-precession.html for more insight on this..

THE GEOMETRY OF UNIVERSAL MIND

stractions and talked of them in terms of gods, their epic feats and relationships.

Modern Society

Thales of Miletus, an astute businessman and thinker, began a system of law that sought to promote reason as a complement to the fire and passion dominating human nature at that time. He was coincidentally enough a teacher of Geometry, and brought left brain analytical precepts to that subject. From that time until recently, science, jurisprudence, and many other expressions of rational mind have dominated in world cultures. Tribal societies gave way to city-states, first in Europe, then North America and much of the remaining world. The arts are deeply influenced by technique, self analysis and rational philosophy. Medicine, sociology and, to a lesser extent religion, now bear the trademarks of a rational age.

It's important to realize, though, that these ages aren't exclusively left- or right-brain eras; one or the other simply predominates. The precessional effect is fairly weak, but over time it establishes itself in each succeeding era, giving it a strong feel.

So if we're to accept the idea of such left-right brain seesawing within Earth cultures, what would be the individual and social traits of each? For any individual in any age, the rational never totally replaces the intuitive, and vice-versa. We are always a mixture of both. A beginning list of traits of each way of viewing the world might be:

The Rational World Of Left Brain

Separation...not of this world

The passive observer – the observer in this world sees the world from a vantage point of separation. He/she is allowed to view the world as if not participating in it. The mechanics of this world are not affected by observation. In the world but not of it.

Details govern

Fragmentation – Thought is trained to view life piecemeal, i.e., the specific, not as a whole. Exclusivity and discernment are the watchwords. The parts of the whole are scrutinized separately. It's not necessary, or even expected, for the parts to create an *orderly* whole. Detail predominates over holism. Science. Math.

Reason is a deliberative process.

Protracted process – Deliberation predominates to make sure all aspects, all viewpoints, are taken into consideration. Process is always trained to make rational, linear steps toward a conclusion. This has the effect of submerging the emotional and non-linear aspects of perception

Order - no mystery

THE GEOMETRY OF UNIVERSAL MIND

and how we make use of those aspects.

Methodology – Planning is crucial. Human activity must be ordered. Mystery is not accepted as a necessary quality of the process. All aspects of process must have the ability to be reproduced consistently.

Isolation - life outside the senses.

Detachment – As parts of the whole are accepted with individual integrity, there's only a rare need to create (or re-create) the whole. I can exist without creating an effect on the world. Human life allows for isolation (so that each part can develop its own integrity). There's a tendency for an individual to live in his/her own mind-derived reality, only scantily clad with sensory input, i.e., in an abstract, two-dimensional version of the world, as if it exists within a picture frame.

The individual rules.

Individuality – The integrity of the part governs over that of the whole. There's the belief that the integrity of the whole is dependent, often totally, on that of the parts. Human power is defined by individual purity and integrity.

Parts determine the whole.

Whole equals the sum of the parts – The value of divisions of the whole, i.e., the parts, determines the identity and integrity of the whole. The whole is always and only composed of an inventory of the parts.

Issues – Social interaction between individuals, because of the push and pull of many individual identities (read: realities), strives toward a workable composite unity, which allows the integrity of the individual to drive the process. Fairness. Equality. Right and wrong. Hierarchy. Power over.

The Intuitive World of Right Brain

In the world - our senses would have us engaged in the world.

The integrated observer – The observer is inescapably a part of this world. No decision, no observation is without effect. It's not possible to be in the world without some level of participation. The world of the senses predominates, touch in particular. Art. Music. Poetry.

Holism – The whole of the world is seen in its entirety, and the observer recognizes that the whole has its own integrity. Because the observer recognizes this, there's no need to search for the integrity of parts of the whole. The integrity of those parts leads directly back to that of the whole. Any division of the whole exists as part of the whole; therefore it has meaning.

Rapidity – Intuition, which naturally recognizes the whole and its built-in significance, isn't concerned with reasoned deliberation. In the face of the intuitive "Aha!" moment, which is instantaneous, method and deliberation seem only to frustrate and unnecessarily slow down the fruits of thinking.

Passion – Emotion comes to the fore. All obstacles to be overcome are reasons to display zeal, love, excitement. Sexual activity tends to epitomize these, and as well seems to be the glue that yields union in intimate relationships.

> Passion and sex

Relationships – Because individuality does exist, its significance is recognized in terms of its relationships with other individuals. Because of common goals and similar identities, such relationships immediately point the way back to the whole. A human individual is defined, not as a solitary phenomenon, but by its mode of interaction with other humans. Power is established, not by its ability to dominate, but by its adaptability and flexibility. Extending human relationships results in the tribe, the society. Within relationships there's an inherent disregard for linearity, and for the *quid pro quo* that makes left-brain relationships work. Rightness is recognized by the feeling and timing of a thing. Time is viewed as cyclic and whispers that nothing is lost because of time; everything worthwhile will come back around. This recognition implies that cyclic time has the ability to balance and therefore to heal.

> Individual defined by interaction with others.

Issues – unchecked emotion and sensuality. Personality and individuality overwhelming others. Cooperation difficulties harming the welfare of the family, tribe, or society. Lack of sense of one's rightful place in the tribe or society.

> Predominations in context

It's important to place these two predominations into proper perspective. This is especially vital in the transition times between rational and intuitive eras. For instance, on early twenty-first century

Leaving left brain.

Leaving right brain.

The dilemma of an age of transition.

The future unfolds...will there be light or darkness?

Earth, we seem to be leaving social expressions of the rational behind and shifting towards the intuitive.

Such times have their own issues. For instance, we spend much of our lives on the flat surface of a book page, a television, computer, or movie screen where reality now demands three dimensionality or beyond. We've become isolated, detached from each other. In a world demanding quicker, more accurate responses to life's difficulties, we're moving slowly, methodically. We become mesmerized by details and struggle to see or understand the "big picture." We find ourselves ignoring many aspects of bodily life, believing that, from our sense of detachment, existence there is unimportant. With this baggage we move toward a world that demands holism. Our intellect wants to ignore the senses. We become more detached, this time in an unbalanced way. The external world becomes less real. We drive ourselves even deeper into our individuality and away from its balance with the human needs of the larger societal whole. We become confused, trusting neither our intellect nor our intuitive skills. We become progressively more mistrustful and cautious in our interactions with others. We cannot quite commit to anything or anybody; we find it hard to resolve the ensuing dichotomies. Personally and societally, human life suffers.

For one leaving the world of right brain predomination to that of left brain, there are similar issues. Accomplishment through our use of passion and emotion is frustrated – the world moves too slowly. Proper attention isn't paid to detail, and failure quickens. Slowly, the significnce of the individual blossoms. But this individual quickly finds him/herself confronted with personal moral issues – he/she deems oneself socially worthy or unworthy. Either way there is an urge to dominate others.

If both categories of these social dilemmas seem innate to our age, it's with good reason. As the old predomination wanes, the new one not yet fully established, those individuals favoring one predomination or the other will find no social structure they can depend on. Solace in such times is in finding balance between the natures of both left and right brain cultures.

If the people of the world can, personally and collectively, balance these two natures, a healthy movement into the next predomination can happen. These two predispositions are a pair of nature's myriad complementarities. The balancing of them in such times affords a rare opportunity for personal synthesis and movement to

Possibilities...

higher degrees of happiness and freedom. In society, such a balance affords an upward spiral of human progression and development. In these times of transition, it's always a human choice to allow this balance to emerge. If we choose to, the individual moves toward fulfillment and society prospers. If we choose not to, the growth of society and the individual slows. The effect is a growing spiritual darkness.

But what of conceptuality, of Geometry? A choice of balance affords a choice of evenness between the random manifest creation and the meta-physical conceptual, which orders the manifest creation. When this balance is struck, there's awareness, there's enlightenment, there is freedom to play among the polarities of created life.

Who knows what opportunities exist for a humanity that has consciously chosen to balance these seemingly opposing natures? Dare we even consider the creation of a truly new reality, freed from the constant fluctuation between the different, if complementary, expressions of mind? Do we dare to think that the whole of created reality is ours to nudge in a new, higher direction? Perhaps by embracing these complementarities, by creating our balance points between them we shift toward possibilities we can now only imagine.

1.3 Exercises

To facilitate the balance between left and right brain, try these:
• With a pencil or pen begin connecting the "+" marks of Figure 3 with straight lines. Allow the mind to freely choose the points and directions. Don't plan your moves.
• As the page fills with lines, look for shapes, pictures (See figures 4-7). Use your imagination. Lines can be drawn from "+" marks to intermediate points on the lines if something begins to emerge.
• If you see Geometric-like shapes predominantly, you favor left brain. If you see pictures, i.e., faces, scenes, etc. you're favoring right brain.
• Repeat the exercise until you can see Geometric shapes and pictures with equal facility.
• It's an interesting point that each person doing this exercise starts with the same matrix of "+" marks. Your unique nature, interactive with the matrix, creates unique shapes and pictures different from anyone else's.
• Draw Geometric shapes or play with Geometric models while listening to delightful musical recordings.

Figure 3

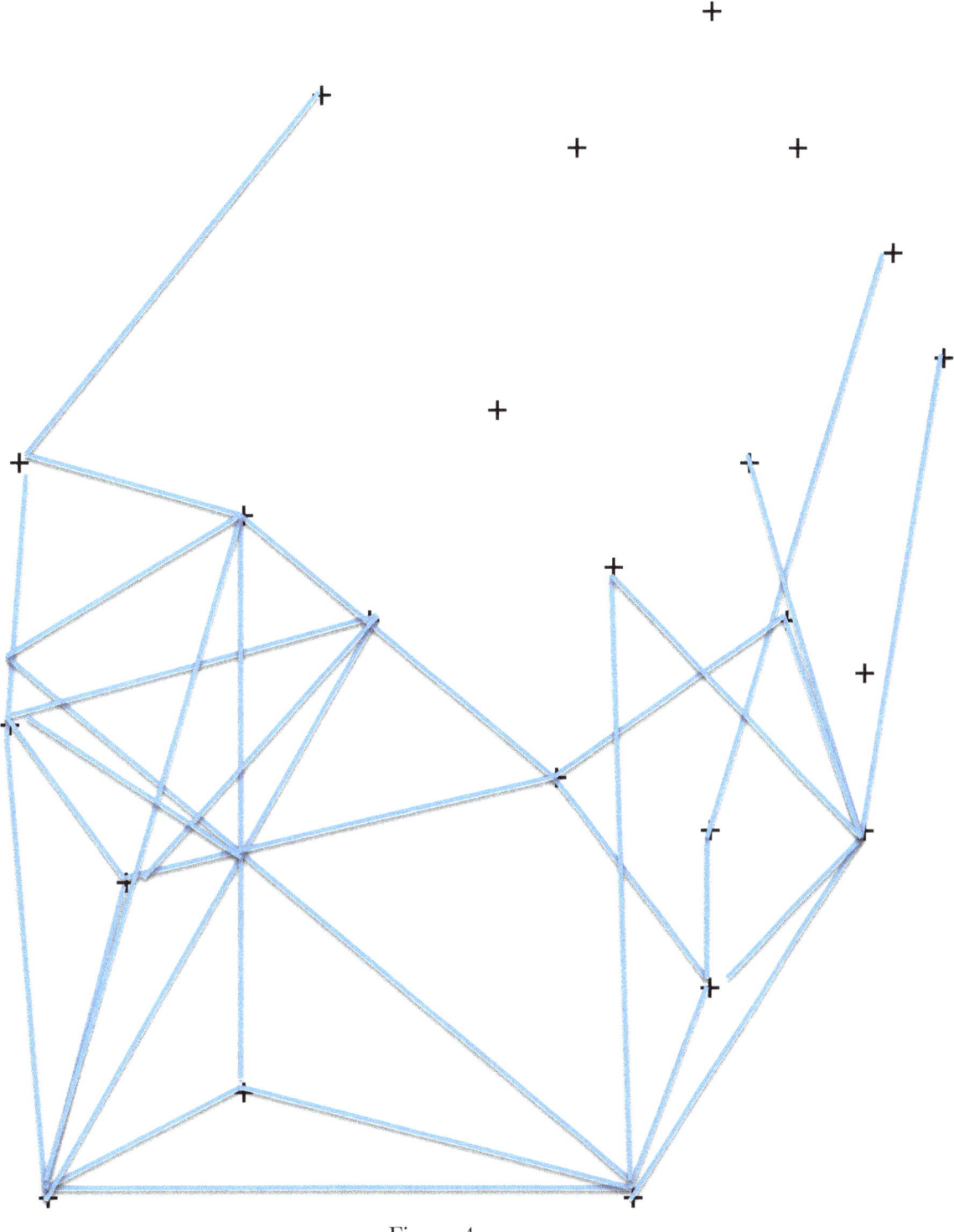
Figure 4

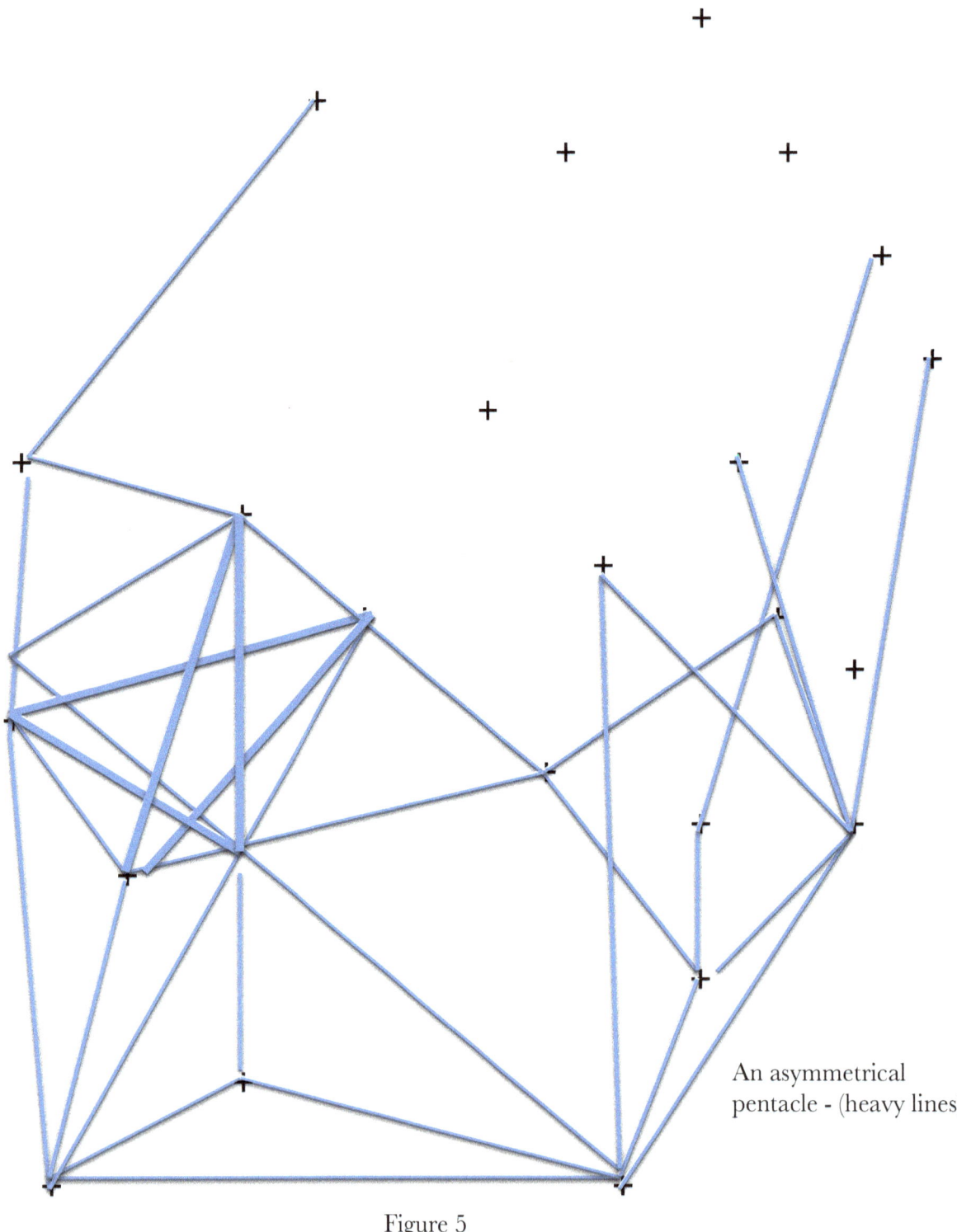

An asymmetrical pentacle - (heavy lines)

Figure 5

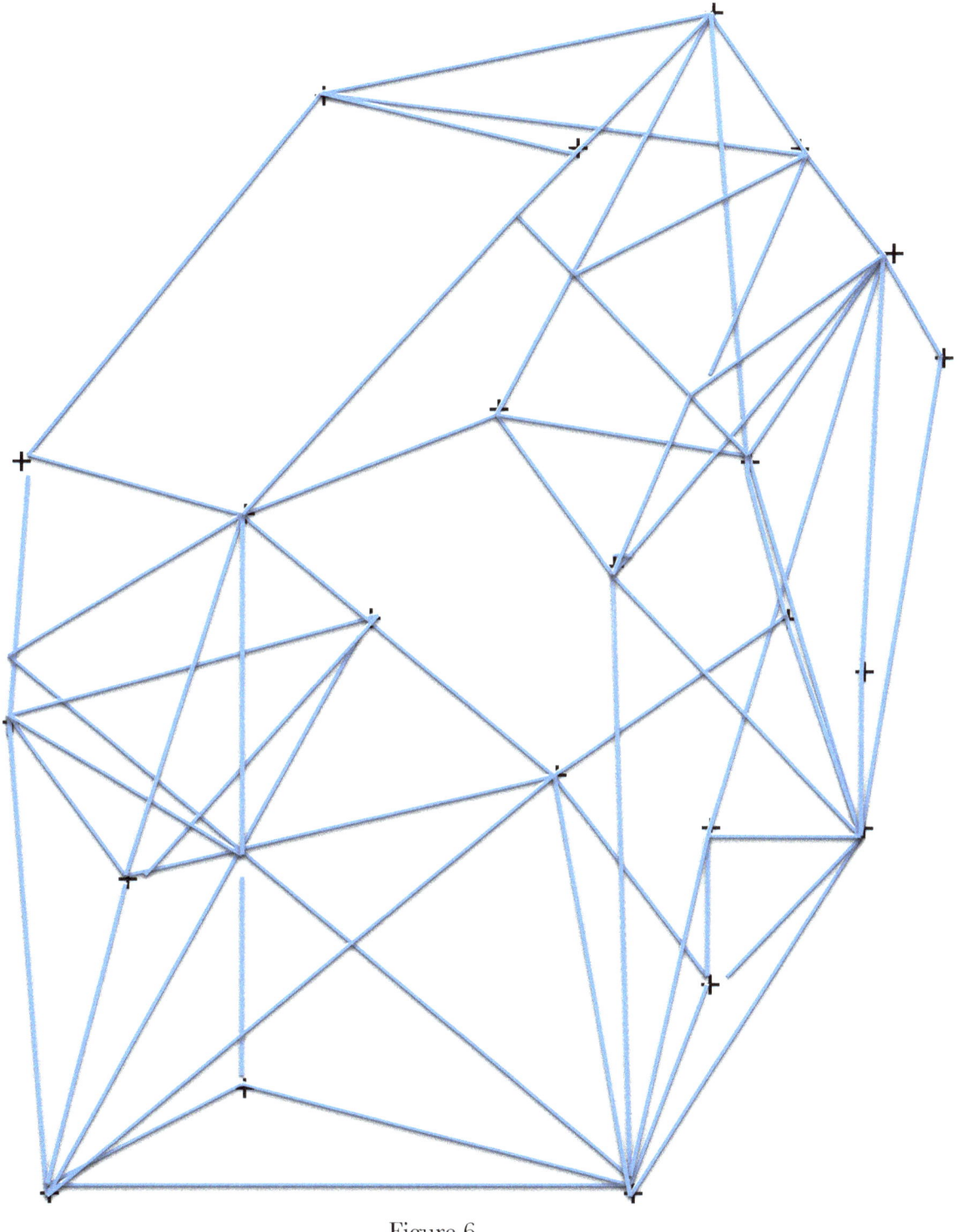

Figure 6

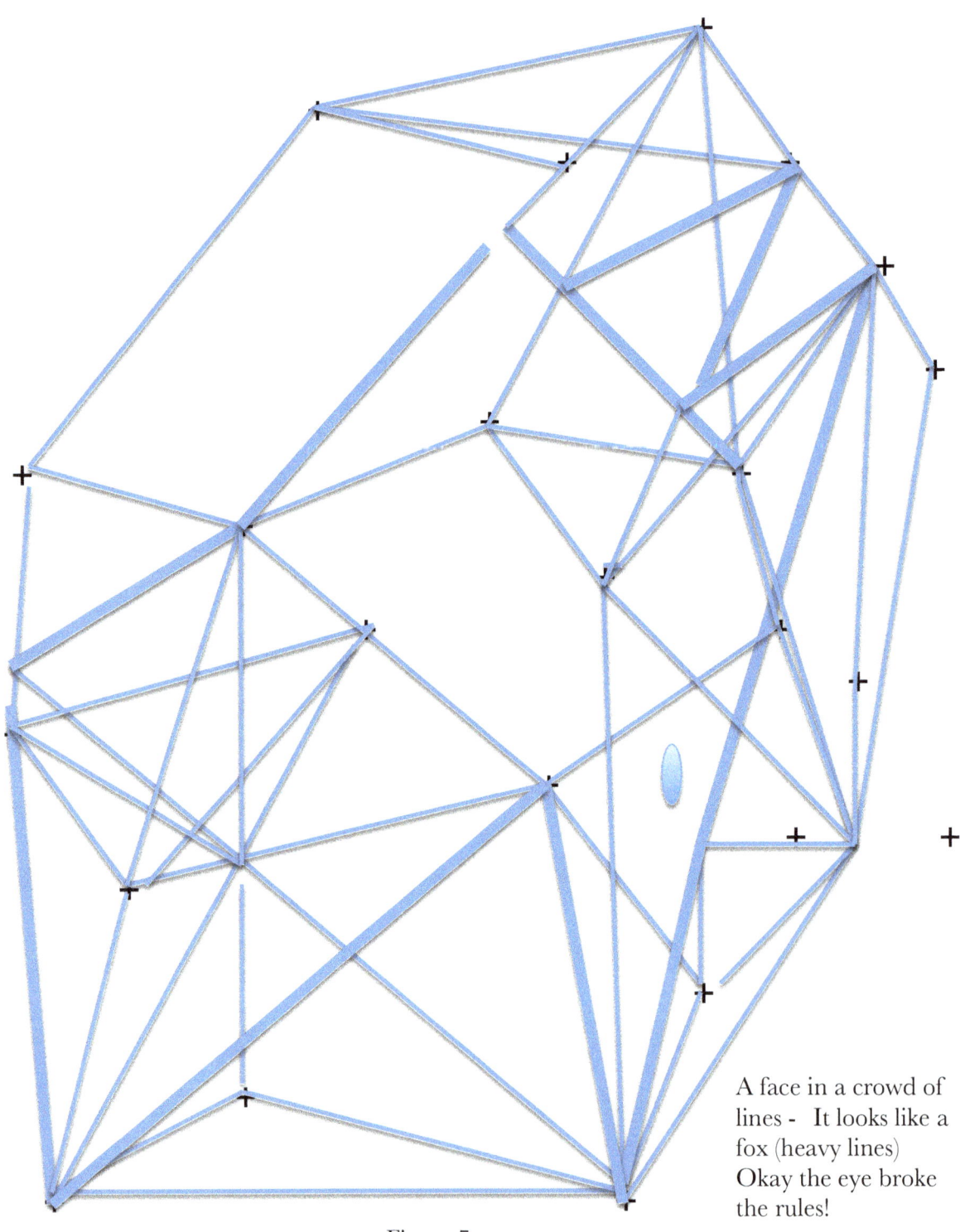

A face in a crowd of lines - It looks like a fox (heavy lines) Okay the eye broke the rules!

Figure 7

The History
and
the
Tradition

2.0 The History of Geometry

This section will focus on the way Geometry has worked its way into humanity's expressions of Mind, i.e., how humanity has socialized Geometry inasfar as we have indications of it in historical records. Some of these expressions will simply be visual/artistic shapes. Others will be mathematical, and still others will be rather philosophical, indicating that Geometry of various sorts has been used consistently throughout history, and to a degree in prehistory, to help understand our reality. These things will prepare the reader nicely for the more modern expressions of the Geometry of Universal Mind in a future volume of this work.

The cycles of Mind, as proposed in the previous section, alternating between the intuitive and the rational, have surely affected how human mind has made use of Geometry, and where possible in this section these alternations will be highlighted. Though history may eventually recognize Geometry as a universal attractor of sorts, it's difficult to inventory a progression of Geometry's role in human life in a comprehensive way. So the following section will only try to provide a thumbnail sketch of the Geometry tumbled historically from our collective thoughts. Where possible, though, this section will try to lay the groundwork for what may very well be the ultimate patterning of Geometry, a patterning embracing the symbols we already know so well.

The experience of twentieth and early twenty-first century humanity has been to greatly expand its direct knowledge of the created Universe. Although that experience may seem wonderfully unique, it's likely that our discoveries and experiences have been there before in other cycles within Universal Mind. The writings in the last century of Hindu seer Sri Yukteswar proclaimed celestial cycles such as our sun's revolution about a dual star (a 24,000 year cycle), and that tandem's rotation about a galactic center, which may also affect how humanity has experienced these cycles. Astrophysical research may ultimately provide a picture of human experience as built on a complex of these, creating a multidimensional playground for human mind. And within this eternally recurring playground we may discover windows of opportunity – windows allowing humanity to realize itself as a part of something grander, reintegrated with the primal wholeness of Universal Mind, and enhancing that primal wholeness' basic structure.

But you may be asking yourself: Wouldn't it be arrogant to consider human enhancement of something so basic? I think not. To the contrary, wouldn't humanity's broadening experience and understanding make such modulation

---

The evolving expressions of Geometry

Other cycles which might have affected mind.

A possible picture of expressed human mind

Reproduction of a spiral configuration by Stone Age tribes of Britain

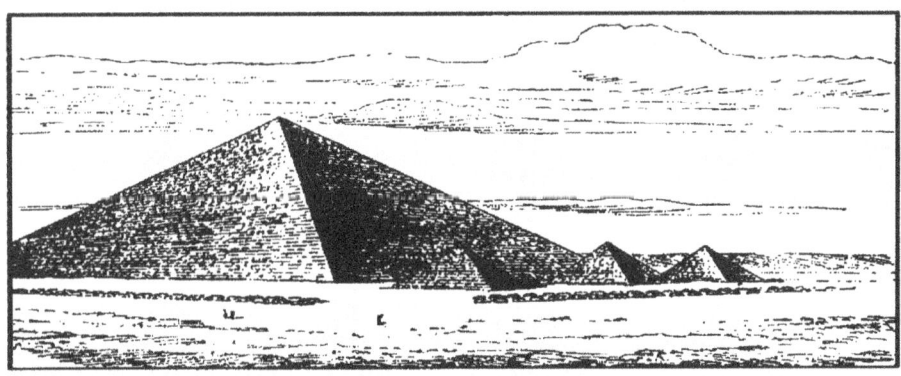

Octahedral-shaped pyramids of Egypt

Celtic artisan design - note the ornate and opposing curves and spirals

Figure 8

Octagonal Greek design

Use of the octagon in Roman design

The use of Geometry in Greek and Roman art-----
Notice the choice of juxtaposition giving the Greek a more "artistic" flair. The Roman, accented by the background brick gives the more "linear" impression despite its spiral border

Figure 9

of Mind's structure inevitable, maybe even desirable? Could it possibly be that such modification would further perfect the Universe's structure?

So back to history. Among the artifacts humanity has left us, we've always found bits and pieces of Geometry. Europeans of the Neolithic era left stones carved in the shape of traditional platonic solids. Earthen mounds of pyramidal shape remain in the Americas. In China. In the Middle East. In Southeast Asia. Spiral shapes are found on large boulders in rural Britain, carved by peoples dating to antiquity. The Vedic cultures of India and Southeast Asia, their origins lost in countless cycles of the past, leave us the ornate Geometric shapes of their yantras, probably perceived during deep meditative explorations of Mind. Baskets and other woven implements remain from the Anasazi of North America, incorporating many complex Geometric patterns. In other words, Geometric shapes from virtually all indigenous cultures of Earth have been handed down from the beginnings of time.

We're most comfortable in our own explorations of history reaching to the earliest Egyptian cultures of the sixth millennium BCE. During this epoch we find the first hieroglyphic writings etched on the monuments of those cultures, these structures usually squared off and tapered shapes rising upward. As that age gave birth to the sciences and architecture of Egypt's Old Kingdom, the famous pyramidal shapes appeared. Temples, palaces, and other structures made the first obvious use of Geometric proportions and ratios. The later Egyptian Middle Kingdom ensued as time cycled back to right brain expressions, leaving us its great art, literature, and artisanship.

An explosion of other, similar cultures appeared in and just prior to the second millennium BCE. The Sumerians, with their pantheon of gods, their mythology, and epic literature, left us ornate circular patterns reminiscent of floral shapes. The Babylonians constructed their squared, pyramidal ziggurats, incorporating stairways in triangular and spiral-like configurations.

From the right brain predominance of the New Kingdom of Egypt, the mysterious kingdom of Sumer or Sumeria, of Babylonia and the early Greek, came intuitive precepts emphasizing dynamic versions of physical Geometric shapes. These Geometries received names; in their archetypal forms they were considered to be gods. They were recognized as fundamental, causal energies, woven into great cosmic dance movements. From such Geometric divinations came the mythologies of those cultures, giving rise to the dramatic epics of their literature, and thereby being woven into the interplay of human life of that time.

And as that artistic age began to give way to the recent Piscean Age, to its explorations of reason and rationality, the Greek and Roman cultures emerged.

Geometric shapes from all world cultures.

See Appendix "C"

Old Kingdom - left brain

Middle Kingdom, others - right brain.

Ancient Earth Symbols
incorporating Geometric shapes

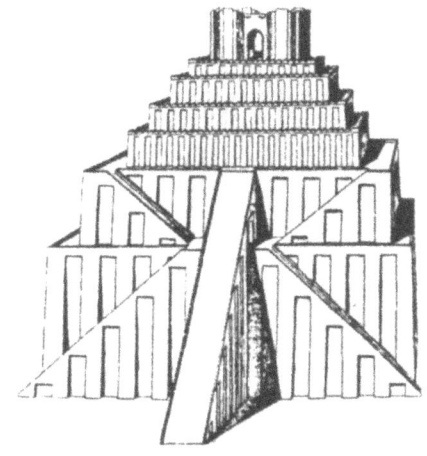

Reconstruction of Babylonian Ziggurat

Native American designs incorporating both linear and non-linear Geometric concepts

Figure 10

THE GEOMETRY OF UNIVERSAL MIND

Right brain expressions: myth, archetypes, art, causal energy

Left brain expressions: order, structure, philosophy, mathematics

Plato synthesizes the abstract and the concrete with Geometry

Number enters the dialogue of Geometry

Influenced as they were by the Egyptian civilizations and the cultures of the Near East and India, they synthesized the grandeur of those cultures' architecture, mathematics, philosophy, art, and astronomy with their own deliberate blends of linear and non-linear Mind.

However, as the at-arms-length nature of left-brain predominance overtook humanity during the height of the Greek, and then the Roman cultures, these individuated, dynamic Geometric shapes settled once more into the abstract eddies of manifest mind. They were no longer mythological and, as such, no longer represented human interplay. They came to represent a sense of absolute order and structure. Stolid Geometric shape, separate from substantive interplay.

Plato saw these new Geometries as the most refined and essential language of reality, hence the ideal language of philosophy. He considered no archetypes or mythologies. Instead he saw Geometry as the language of the soul, something innate in each of us.

From Plato's precepts a sense of process evolved, principally from his Geometric shapes. One could come to understand philosophically the bases of all phenomenal creation. He reached into the antiquities of the Egyptian, Sumerian, and Babylonian cultures for a connection with that age, returning with his five fundamental shapes, all solids: tetrahedron, octahedron, cube, icosahedron, and dodecahedron. These were shapes you could hold in your hand, and their prime significance grew to be a tactile connection with the mathematical abstractions developing during that age. This symbiotic relationship thereby defined the evolving culture of that time, greatly affecting not only the rational thought of its law, architecture, philosophy, and mathematics, but its artistic sensibilities and its social structure.

And so number entered the symbolic language of Geometry. "ONE" came to represent our deeply felt sense of Unity of all things. "TWO" was used to capture the essence (polarity) of opposites that Nature imposed on itself. "THREE" captured the essence of trinity – the three-fold interplay of forces at work in Nature and was seen as the fundamental window through which form would be noticed by the rest of us. As such, "THREE" was termed the Mother, the membrane or veil through which life entered into form. And "FOUR" was the conceptual word for completion, the primal sense of creation settled into the structures of form.

As Pythagoras pronounced, influenced as he was by Thales and his use of reason to survey reality, "all is arranged according to number." So we came to refer in the language of mathematics to "oneness and "twoness" and "threeness" and on and on, as Mind's ever-real sense of interplay began to be superimposed via math on the discrete, fixed structure of solid and plane

A Yantra -
nine interlocking
triangles - four down,
five up.

Japanese calligraphy figures representing "creation":

Figure 11

THE GEOMETRY OF UNIVERSAL MIND

Pythagoras and number

New, yet old, cyclical problems

Thinkers struggle with the demands of rational mind in a time of transition

versions of Geometry.

This age we live in, this twenty-first century, has its challenges, as have all previous eras. In the twentieth, left brain predominance began to wane. We spent our time dealing with the minutiae and extreme details we conjured up from left brain predominance, and we all but ignored the larger picture our philosophies and sciences could give us. And so now we're forced to blink a time or two and see if that composite of our mental endeavors comes into a holistic focus. To see, as the much older versions of philosophy and science discovered, that these bits and pieces of reality could transform, propagate and regenerate themselves.

So in the earlier age, Thales' thinking led to Pythagoras' use of number and rational analysis in coming up with a system of Geometric proofs. The change to rational mind is never all that predictable, though. This led to Anaxoras' efforts to have us understand the cosmos, his commingling of science and revelatory experience. Then came Zeno, who rebutted some of Pythagoras' use of numbers, and a great debate ensued over a new wrinkle: paradox. Principally these debates tried to resolve the supposedly rigid-in-shape-and-size Geometric forms the thinkers of the day thought they understood with the real-life situation of constant change. How, the debaters asked one another, can an object occupy space and yet move through space at the same time? A simple question, no? But one we're only now beginning to answer through relativity and quantum mechanics.

These guys lived, worked, devised, and scratched their heads in perhaps the most difficult kind of time, that of transition between left and right brain predominance. Rational thought came to the fore then, but colored deeply by the intuitive, epic sensibility of the previous age. A lot like the challenges we face today in solving problems left to us by the then-encroaching use of reason, isn't it? Today it's more or less the opposite of that older time: we find reason has made a huge dent in the unknowns of our reality, but it has also left us with certain principles, those of uncertainty, to once again cloud our view.

2 inches

The "root" of the middle square is the diagonal of the one inch square - and one half of the diagonal of the two inch square.

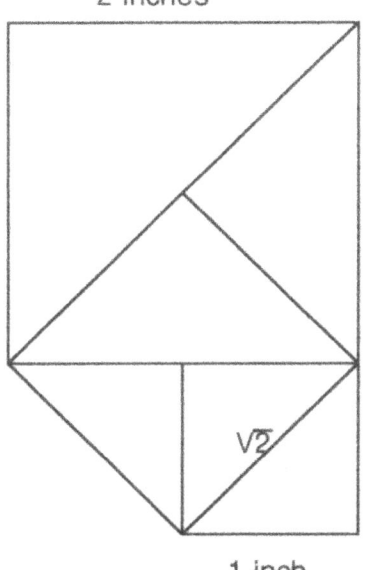

A square grows in size through the multiplying effect of the ratio of the length of the square's side to that of its diagonal, i.e. through the numerical value of the **square root of two.**

1 inch

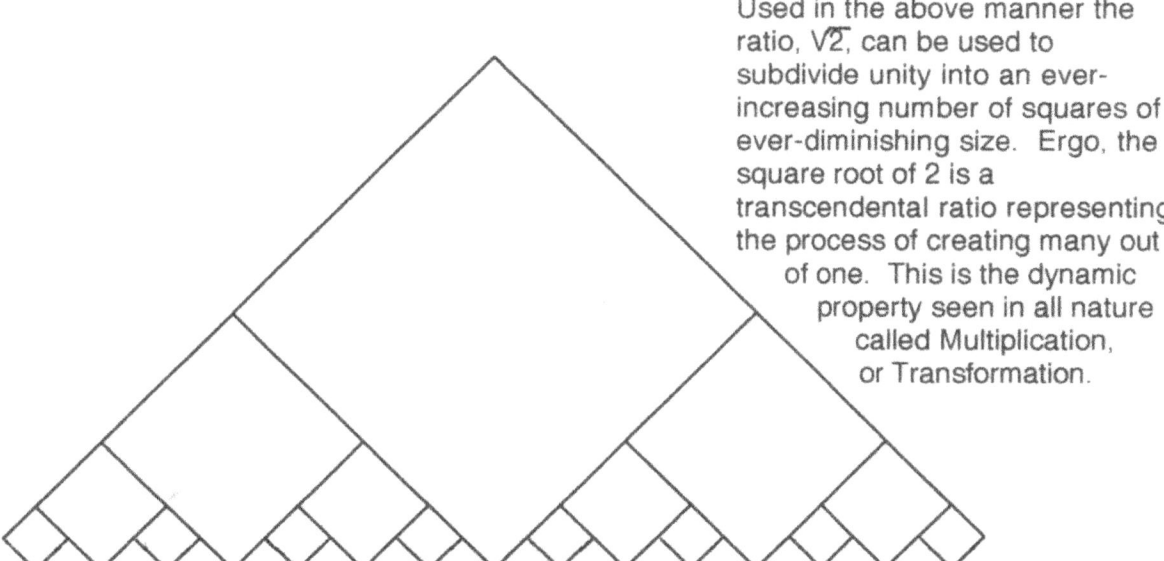

Used in the above manner the ratio, √2, can be used to subdivide unity into an ever-increasing number of squares of ever-diminishing size. Ergo, the square root of 2 is a transcendental ratio representing the process of creating many out of one. This is the dynamic property seen in all nature called Multiplication, or Transformation.

Figure 12

2.1 Geometric Analysis

The Greeks found it necessary to figure out how to represent their era's social friction so they could return it to a sense of basic unity. And they had a relatively new tool – mathematics. Slowly the visual sense of proportion in pursuits such as architecture moved toward that of number in mathematics. And these became the central tools in restoring the jigsaw pieces of their emotional, intuitive life to a grander picture of unified reality. Their strategy was to work backwards in using math to reconstruct Unity. Somehow, they realized, the first moments of the Universe's creation divided its overarching sense of Unity. How in the world could they explain this?

For the Greeks their adventure into logic and math led them to a new symbol for Unity. No longer was Unity a state that the Universe returned to in dynamic cycles. Now the ancient symbol of Unity, the circle, gave way to a growing insistence of the new mathematics that reality was structured in a ninety-degree, perpendicular-angled orientation: the square.

But the square already had a unique symbolism of its own. It was symmetrical; all sides and all angles, internal and external, were equal. The four sides represented the four directions: north, south, east, west. Curiously, in the jargon of this new mathematics, any number that was multiplied by itself was said to be "squared." And so the language of perception was juxtaposed over that of mathematics, with Geometry lying at its basis.

To further develop the metaphor of multiplication, these mathematician-Geometers, began to toy with the line drawn between opposite corners of the square – its diagonal. The sides of the square were termed the diagonal's **roots**. The diagonal was the **length**. Then, using the diagonal of a small square to construct the perpendicular sides of a larger square, they discovered a number of curious things. For a one-inch square generating a larger square, as shown by two "multiplications" of the diagonal (Figure 12), the side of the larger square sitting atop the smaller was double that of the smaller square, or two inches. But there was another square in between, a "diagonal" square. Its *length* was equal to the *root* of the larger, two-inch square. And the *root* of the two-inch square was also equal to the *length* of the first square <u>squared</u>. Looking at it a different way, the *length* of the first square could be set as our original Unity. And the proportion of *root* to *length* still held no matter what the sizes of the squares in question were. So we now had a mathematical **quality** mimicking Nature's habit of multiplying its original Unity.

But they discovered more. With the ratio between the *length*, the diagonal, and the side, or *root*, of any square being the same regardless of the size of the squares, they sensed something absolute about this proportion. In a world increasingly involved with the relativistic nature of linear thinking, the square root of two in the world of squares seemed to transcend the measureable world of form, and even the way the interactions happened.

    So from this doorway into linear mind, abstraction emerged in the form of the square root. Its newly unearthed nature as the link to creating big squares from little ones, little squares from big ones, became the key to subdividing Unity. This phenomenon had been noticed in other ways early in human awareness and, now in the grip of mathematics, this process was termed **Multiplication**. And so this began to answer questions that had cropped up in philosophy: How do human beings, the leaves on the trees, grains of sand, procreate? All by some version of dividing the absolute concepts of "humanness," or "leafness," or "grain-of-sand-ness" by the process of Multiplication. These Geometers and their forebears had recognized that form is always a complex of Geometric shape and configuration. So it wasn't surprising to discover that this multiplying effect, moving toward both unlimited expansion and ever-smaller minification, also represented the cellular growth of living tissue.

    As our Geometers took to a deeper level of inspection and understanding, the process of Multiplication came to be considered in tandem with **Transformation**. It's wrong, they said, to think of any bit of matter, any elemental substance, as the thing it seems to our senses in any given moment. The processes of Nature are always in motion, always on their way from one state to another. Water is always boiling into vapor or freezing into ice. Organic materials are always decomposing, being assimilated by other materials and their processes. Common soil is always in movement from decomposed plant matter and degraded rock to nourishment for other organisms. Static Geometric shapes, such as the square, and the newly evolving relationships between angles and sides, represented an increasingly dynamic, more sophisticated way of looking at Nature's processes. But the great minds behind these groundbreaking thoughts and modelability knew there was more to be done in accurately creating a conceptual model of a dynamic world. Looking still deeper, they knew they had to explain the processes of Multiplication and Transformation, not simply in terms of the division of Unity into identical portions of identical makeup. The assimilative, digestive, transmutative processes seen in Nature began seemingly with one substance and ended with another. On a basic level, one might combine straw and mud and obtain something unexpected: brick. Two gases, hydrogen and oxygen, might combine to form a fluid, water. How was this possible from the standpoint of their evolving conceptual models?

    Geometrically, their thinking progressed from the plane, flat, unreal world

---

Multiplication as a concept

Multiplication at a deeper level = Transformation

The cube

of the square to that of the cube. A square can only exist as an idea, perhaps represented by a symbol on a piece of paper. A cube, however, is a real thing. It exists in crystalline form in Nature and can be fashioned by human hands from stone or clay or wood. The cube is a substance you can hold in your hand, its shape defined by the idea of interlocking squares oriented in space at ninety-degree angles, all equidistant from another. Upon examination through the evolving mathematics of the day, our Geometers discovered the diagonal of the cube. Like that of the square, there was a constant proportional relationship., or ratio, between the length of this diagonal and the all-sides-equal lengths of the cube's edges. As in Figure 13, a cube could be constructed with edge lengths equal to Unity. A diagonal of any square facet of the cube was $\sqrt{2}$. Three points define a plane and, incorporating the known lengths of AB and BC, the length of the cube's diagonal, AC, could be calculated. The diagonal length of our cube was, by Pythagoras' rules of squaring sides, the square root of three. And once again, the proportional length of AC and the cube's edge length, CB, were found to be constant, independent of actual side length.

Cube and Square root of three

There were several philosophical implications concerning the square root of three to be discovered. First, if the square root of two represented the rudimentary division of Unity, or Multiplication, our new concept represented extending the analogy to accommodate a division into unlike parts. The cube, as a shape that can actually be created in form, took the relationship off the surface and allowed it to accommodate a sense of depth. **Depth connotes more than spatial volume; it implies that it's possible for substance to change as well as shape.** And in doing so it says that by using the multiplicative process in a manner in which the original substance is allowed to transform, material form is allowed to occur. In other words, through a process of shape and substance change, which is eternally dynamic, material reality comes into form.

Shape and substance

Meanwhile some of our Geometer friends returned to the circle as a symbol of unity, and they made similar discoveries. They allowed two circles to overlap such that the center point of each lay on a point of the other's circumference. A common area was created within the overlap, which was imagined to be in the shape of the cross section of a fish. In the centuries to follow, this inner fish became known as the ***vesica piscis***, or the bladder of a fish, connoting an organ of transformation.

Vesica Piscis

Within this interior shape they constructed perpendicular lines: one connecting the centers of the two circles, the other connecting the upper and lower intersect points. This construct, with four cardinal points, connected its interior message with that of the square and the cube, thus giving their Geometric explorations a powerful connectivity.

1 inch square

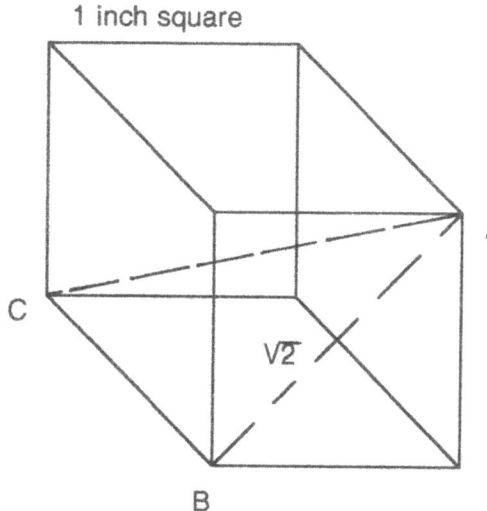

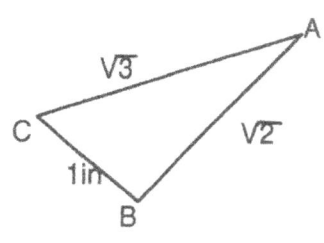

From Pythagoras:
$BC^2 + AB^2 = AC^2$
$1 + 2 = 3 = AC^2$
Therefore $AC = \sqrt{3}$

All sides of the two squares intercepting the overlapping circles that create the *vesica piscis* set equal to Unity, or 1.

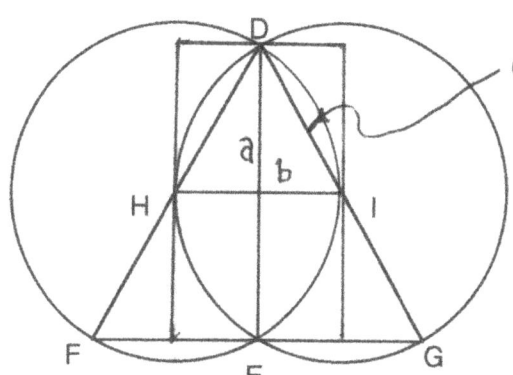

From Pythagoras $a^2 + b^2 = c^2$
Where c = Unity = 1 and
$b = 1/2\,c$
$a^2 = c^2 - b^2$
$a = \sqrt{1^2 - (1/2)^2} = \sqrt{1 - 1/4}$
$a = \sqrt{3/4} = \sqrt{3}/2$
$2a = \sqrt{3}$

The Greeks' original construct was to connect HI and DE, creating two perpendicular lines and four cardinal points. Alas, they didn't discover the triangle DFG, which also contains their four cardinal points.

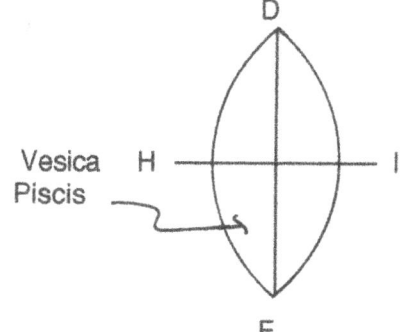

Vesica Piscis

Within the Vesica Piscis, if HI = Unity, then DE = $\sqrt{3}$

Figure 13

As an aside, it's generally reckoned that this construct played a vital part in the development of the Greeks' evolving belief that the ninety degree, or perpendicular, orientation was at the basis of nature's Geometric structure. Had they performed an equally impeccable construct by connecting the upper crossing with each circle center, they would have discovered the sixty-degree orientation of the equilateral triangle. Alas, their use of this orientation would have resolved much of the numerical irrationality in the mathematics of subsequent years.

Vesica Piscis and Christ

The association of this shape with the fish was also adopted by Christian mystics in their own use of Geometry to explain reality. With Christ as the centerpiece of their philosophy, this fishy shape became the symbol of that-which-joins-together. One of the mystical messages connected with Christ was that of a deeper penetration of spirit into form. Therefore, the *vesica piscis* was the common ground, the place where the unmanifest coexisted with form. A conceptual place in which duality coexisted with a hint of Unity. It was thereby a cosmic resolution: it was neither spirit nor form. It was, in fact, something new, which was neither and both–it was their synthesis. In truth, this definition was an excellent identification of the Greeks' Transformation. And coincidentally, the discovery, naming, and use of the *vesica piscis* corresponded to the advent of the astrological Piscean Age.

But there was more to be revealed from the emerging superposition of conceptuality and mathematics upon Geometric shape. While Multiplication and its refinement, Transformation, successfully represented the processes occurring within Nature in which material substance moved, reshaped, restructured itself, it remained to discover the way of life and death. The thinkers of the day knew that Spirit finds a body; it lives there and then departs. Trees, animals, homes, nations come into existence and then go away. But beyond this, there was the historically overwhelming consensus that this alternation between being and seemingly not-being was part of a grander phenomenon. A phenomenon in which life was eternal, no matter the phases of form created by that life. Surely, these persons considered, there must be some way of representing this alternating between life and death, together with its underlying sense of eternity, by Geometric construct.

Life, death and Eternity?

And they were right. As they looked deeper into the complexities of the Geometries, they discovered another ratio, another conceptually independent proportion. And upon this one they built their own created world as they discovered its omnipresence in the creations of Nature.

Similarly to the two linear relationships of the square (*root* and *length*) and the two interlocking circles of the vesica piscis, our Geometers came to allow two adjacent squares to represent the seemingly disparate realities of spirit and

THE GEOMETRY OF UNIVERSAL MIND

See Figure 14

form, of stability and flow. As they traversed the square and cube, a diagonal moved across these two squares, crossing their common line at midpoint, to the opposite vertex of the two. Once again, from the relationships of Pythagoras' math, the proportional length of this diagonal with respect to the Unity-length of a member square's sides, was discovered to be the **square root of five**. Much as the square root of two was a conceptual juncture point of the Multiplication process and the square root of three established the common ground of substance change as Transformation, the square root of five provided the doorway for crossing between the worlds of spirit and form. It confirmed, in the most comprehensive language possible, the world of complementary phenomena. Opposites are created to play complementary roles, leading to their synthesis, and then to the over-the-horizon sense of Unity.

Square root of five traversing spirit and form

Once again, the straight lines and conceptual relationships of the square were paralleled by the curvilinear, harder to establish relationships with arcs of the circle. Segments of circles were superimposed on the same pair of squares in a manner defining the five-sided symmetry of the pentagon. In this construct, in which the $\sqrt{5}$ lies between Y and O, the points of multiple arcs intersecting define this shape, also revealing the transcendental proportionality of $\sqrt{5}$ within its bowels (CD:YD).

The Pentagon

These three square root proportions, then, came to represent the acknowledged spectrum of creation in all its manifestations. Multiplicative power subdivided Unity, and through Transformative processes Nature was allowed to become a holistic, Regenerative organism, allowing life to traverse the ephemeral, created world and its eternal complement. These three were linked together in one final piece of symbolism by establishing the 3:4:5 triangle. This three-way relationship allowed the circle, the square, and finally the pentagon their interplay in creating the surface definitions of Plato's five fundamental Geometric shapes (See Appendix "D").

The 3:4:5 triangle

Buried within this symbolic language was a dynamic process, only dimly seen in the linear, surface-mapping world of that age, which somehow accomplished making creation real by **alternation**. As rational mind came to the fore in later centuries to search for the one fundamental quantum of the Universe, the sense of alternation between complementary phenomena became ever more realistic, ever more fundamental. But briefly, alternation in that era was defined by allowing the side lengths of Plato's five geometric shapes (and later, other more complex shapes) to equal Unity. Then, by lopping off (truncating) vertices on these solid shapes, other solid shapes were formed from the original. The number crunchers among out Geometers could then, through their growing skills with math, derive other "numerical" quanta to assign to edges of the resulting Geometric shapes. They could then perceive the lopped

A growing sense of alternation between complementarities.

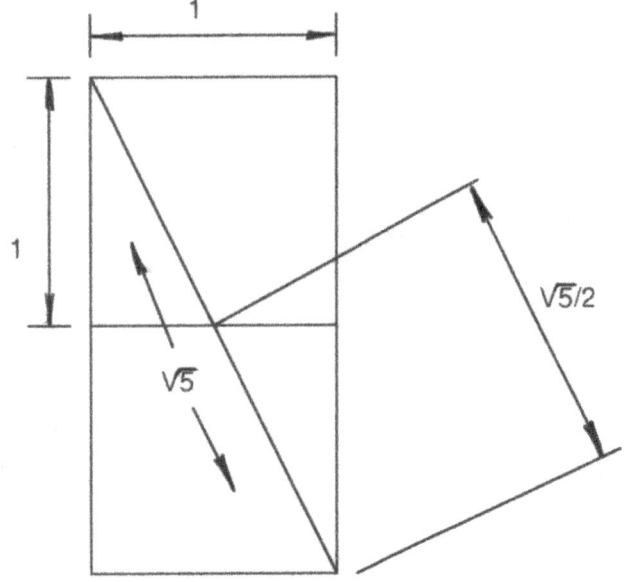

The square root of five traverses two squares as their common diagonal.

Again from Pythagoras:

$$a^2 + b^2 = c^2.$$

$$3^2 + 4^2 = c^2 = 25$$

then $c = \sqrt{25} = 5$

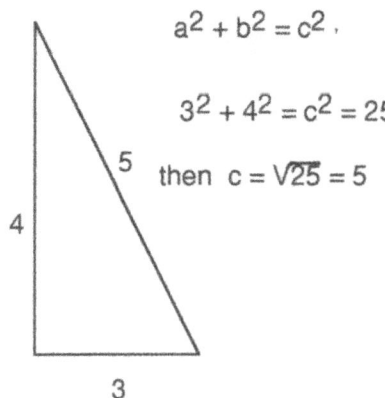

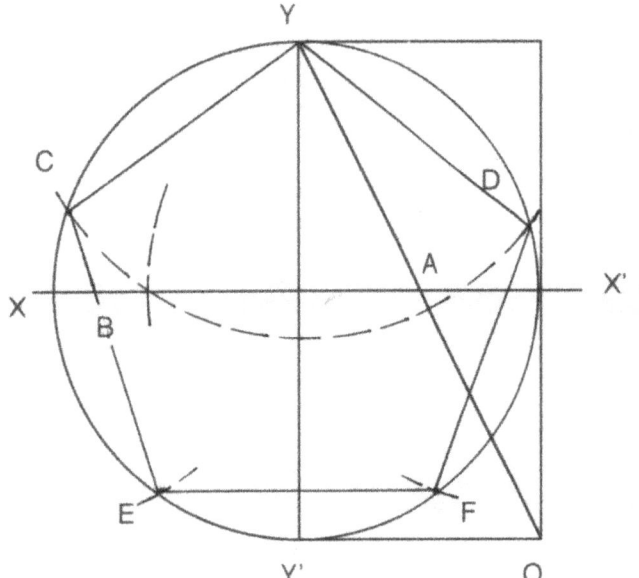

The Pentagon

Using the two adjacent squares of the square root of five and arcs of circles to construct a pentagon. If the pentagon's sides are Unity, the chords, CD for example, are a proportion called the Golden Mean, $\sqrt{5}+1\ /2$.

Figure 14

off pieces, as well as the resulting Geometric shapes, as additive and multiplicative growth. Not an easy thing to describe or envision. But it was a forerunner of the sense of dynamic quanta to be more fully defined by twentieth century science.

2.2 The Golden Mean

So the proportions defined by the square roots of two, three, and five, and the natural processes of Multiplication, Transformation, and Regeneration were at this point represented by numerical values. But there was a missing piece to Nature's creative process that hadn't been accounted for. That piece was the perceiver of these processes. If we return to the original thought of this work, that everything is constructed of Universal Mind, we have to ask ourselves two questions: Who is perceiving what? and What is the process by which this perceiving occurs?

What's the nature of perception?

While the philosophy, mathematics, and Geometric constructs that were debated and defined at the beginning of the Piscean Age may not have overtly posed these questions in quite this way, you can bet they were implied. The observer, or perceiver that they noticed was of the detached variety, characteristic of left-brain predomination. While they sensed in a variety of paradox-laden ways that Universal Mind had indeed devised a manner of observing itself, their observer was conceived to be somehow separate from that which was observed. But no matter; they were successful in defining the process of perception.

Our Geometer friends unearthed the concepts of the knower, the known, and the process of knowing from cultures that preceded theirs, one of which was the ancient Vedic culture centered in India. While the tone of the Piscean Age in the Western World was not the holistic Vedic one, our thinkers did recognize perception's three basic components and the interplay of perception with the observable, no matter how detached.

From the exercises at the end of the first section of this book, you surely recognize that an observer can connect dots in any manner appealing to him/her. This connecting-dots exercise will likely be unique to each observer, even though the collection of dots is common for all and which at the outset have no real significance. But as the various observers begin to draw lines connecting the dots, they may begin to perceive faces, animals, objects, and basic or complicated geometric shapes. Having connected all the dots to their satisfaction, our observers may characterize what they've drawn as either right or left-brain images. (If you see animals, landscapes, faces, etc., for instance, you

THE GEOMETRY OF UNIVERSAL MIND

can say you're viewing the dots and lines through a right-brain filter. If you see Geometric shapes or structures in which such shapes stand out among the rest, then it's left-brain.) So we can say that what's occurring is some cognitive preference for either left or right-brain on the part of the observer, or in this case the line-drawer. **In other words, the persons doing the perceiving encounter a field common to each of them, and they project a whole host of unique perspectives on it.** From the Vedic way of articulating this, the knower, through the process of knowing, projects the inner seeds of its cognitive preference on the dotted paper to create the knowable or known.

Knower/known/process of knowing

But back to the Greeks. As they developed their Geometrically oriented philosophy by this same three-way process, they seemed to want to define the process and its interactions through mathematics. It was easy enough to establish a relationship between two things. Having done that, they took on the challenge of doing the same thing in non-Geometrical math by adding a third thing, as they had done with the three transcendental square root proportions. But to do so took a further stretch of the newly redeveloping rational mind.

A common field/unique perspectives

One item, **a**, could be compared directly to **b**, and the term, **a**, could be compared in the same fashion with a third, **c**. Too, **b** could be compared to **c** in the same fashion. And so how could these relationships be interlinked in the evolving language of mathematics?

If, the Greeks thought, a dependable proportional relationship could be established between two things, it might be possible to say that either of those items could be related to a third thing by the same or a similar proportion. In other words, if **a** is proportional to **b** in a certain way, then that proportional relationship, or constant, could be used to compare **b** to **c**. Or:

a:b::b:c, i.e., **a** is proportional to **b** in the same way that **b** is to **c**.

Indirect proportions: a:b::b:c

They proved this to be dependable and, as **a** was successfully compared by this indirect way to **c**, the concept of ratio or proportion gained traction. The concept of subdividing Unity by these mathematical means (commonly attributed to Greek thinker Thales) along with this new method of proportioning, a mathematical version of a proportion long understood to occur unendingly in Nature was devised. That proportion became known as the **Golden Mean**, and its mathematical expression was $(\sqrt{5}+1)/2$. An example form of its development is shown in Figure 16. The Golden Mean contained within it the quality of being able to create physical proportions that enhanced beauty and balance. It occurred indirectly as well in the numerical branching systems of plant life, in the functionality of the human body and in the proportions of certain Geometric shapes.

Nature's proportion - the Golden Mean.

A Mathematical Subdivision
of Unity

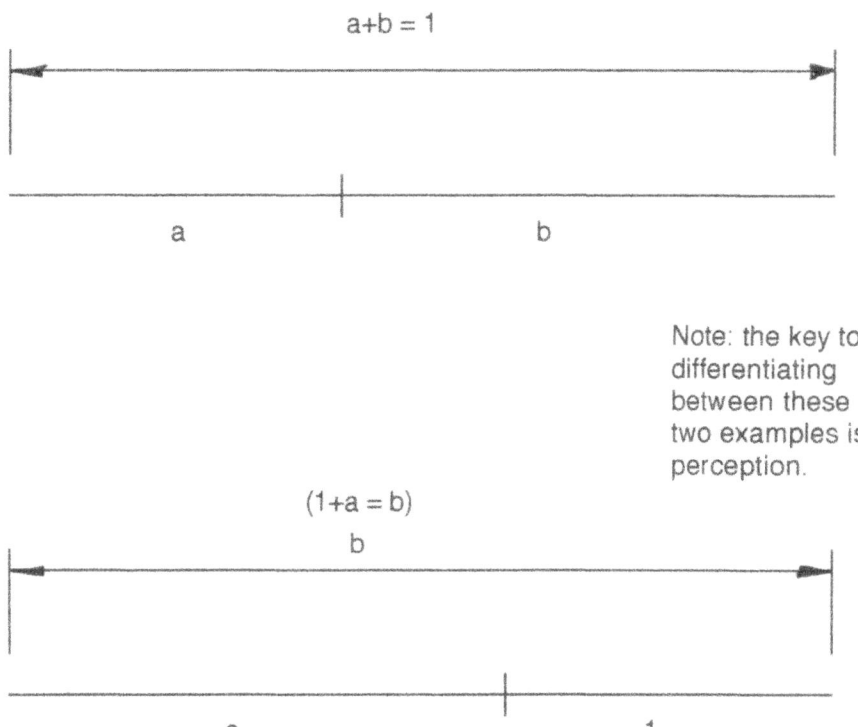

Unity, represented by the number 1 is seen to be multiplied through the process of minification. The whole is Unity, consisting of its subtractive subdivided parts.

Note: the key to differentiating between these two examples is perception.

In this second relationship (above), Unity, or the number 1, is seen to be multiplied by the process of propagation. Unity is the original component by which the whole grows to incorporate its additive, multiplicative parts.

Figure 15

THE GEOMETRY OF UNIVERSAL MIND

The three square roots define a "window" into the created world.

And so the greatest minds of that age determined through their evolving mathematics and Geometric constructs that, through a three-aspect process of perception, a series of transcendental proportions evolved rationally. From the "window" opened by these three proportional constants, the process of proportion then helped acknowledge a sense of created reality. Perception, then, was the device by which conceptual Geometric proportion was given life and was, therefore, dragged into meaning. Perception was the trigger by which life force commixed with conceptuality, creating the whole matrix of sensory experience.

While the role of perception in understanding natural processes was to become more apparent in the centuries to follow, at this time it remained somewhat obscure. Its effect was implied, however, in mathematics involving the Golden Mean. But because perception, like the concept of time, was assumed to be a constant, it drew very little attention to its own unique nature and role in the creation process.

Beyond the uses the Greeks and Romans later made of the beauty defining proportions of the Golden Mean, a philosophical connotation began to revolve around both its properties and its pervasive presence in the structuring of natural processes. The relationship of **a** to Unity, evolving from the constructs shown in Figure 16 was found to relate to the Golden Mean. Its numerical value, ϕ, i.e. $(\sqrt{5}+1)/2$, equals 1.618034 and was substituted for **a** in this relationship (see below). Units of increasing power values could be added to one another, both as whole numbers and as inverse numbers (one divided by the numbers) ++, as indicated in Figure 17, to increase the precision. Also, upon deeper examination of these numerical values, a string of constant relationships could be established between the units, all equaling 1.618 for whole numbers and 0.618 for inverse numbers. This somehow implied an ability of the Golden Mean to multiply "inwardly" as well as "outwardly," or along with higher powering. Gradually the Golden Mean was accepted as a concept so consistently pervasive as to be seen in the creation and processes of all natural phenomena.

Now the Golden Mean, ϕ, is substituted for a to form a new relationship

Hence all examples of the reoccurrence of this constant, numerical value within the Golden Mean were significant. **The additive aspect of increasingly powered values of ϕ in the Fibonacci Series implies the ability of the Universe to divide its Unity in a multiform manner.** They indicate numerical subdivisions, i.e. the making of two things from one and of four things from two. **But the accompanying powering was seen intuitively to represent a quality of the original Unity (ϕ) that did not progressively distance itself from that Unity.** Each powered element represented a subdivision of Unity, which, while progressively subdi-

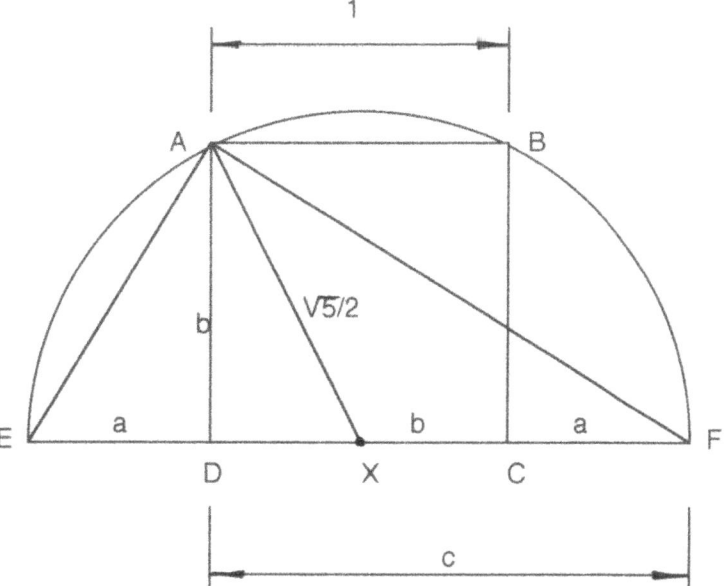

A square is placed within a semicircle symmetrically, i.e. so that
ED=CF=a
AX= half of √5
(See Fig. 14)

From an inspection of the above construct, c = a+b

The Theorem of Thales states that any triangle constructed within a semicircle with each vertex touching the semicircle's perimeter, will be a right triangle, meaning that angle EAF = 90 degrees, as are angles EDA and ADF. Therefore:

    a:b::b: a+b   (EAD and DAF are similar right triangles)

  b=1,   a=√5/2 - 1/2,   c=(√5+1)/2 --- the Golden Mean

since triangles DAF and DEA are similar by Geometric definition, then proportionally,

$$a/b = b/(a+b) \quad \text{then}$$
$$b^2 = a^2 + ab \quad \text{Since } b=1,$$
$$1 = a^2 + a$$

Unity, or 1, has been resolved from a threeway relationship (a, b and c, or a+b) in terms of a single value, a. However Unity is composed of various mathematical forms of a, in this case a^2 and an additive a. In this construct we see the subdivision of Unity by the process of multiplication by propagation. If, as in Figure 15, this manner of subdivision of Unity is seen to be identical with that of multiplication by minification, differing only through the tools and intent of perception, then it can also be said that:

$$1 = 1/a^2 + 1/a$$

Figure 16

THE GEOMETRY OF UNIVERSAL MIND

Additive values imply subdivision; powered values indicate qualities of Unity.

vided in a physical sense, always contained qualities that made it autonomous. That is, it contained the qualities of the original Unity that allowed it to be whole in itself. Each powered unit was then a little Universe, with the power to multiply, to transform, and to regenerate. On the inverse-number side of the ledger, these properties are represented in reverse. Meaning that while the powered numbers represent this process, moving by way of propagation, the inverse numbers represent its complement, which we term minification (original unity growing "more populated" by subdividing itself from within into ever smaller units.)

The mathematics of the Golden Mean and its values of $\phi = 1.618$ and $1/\phi = 0.618$ thus defined a new level of conceptuality far more sophisticated than that of the initial square root relationships. And it was subsequently discovered that the significance of the Mean's relationships could be extended even further by allowing its powered values to be loosened so that they were not additive, but proportional: $1/\phi4 : 1/\phi3 :: 1/\phi3 : 1/\phi2 :: 1/\phi2 : 1/\phi :: 1/\phi : 1 :: 1 : \phi :: \phi : \phi2 :: \phi2 : \phi3 :: \phi3 : \phi4$

The possibility of Universes within Universes.

This expression added flexibility to the Golden Mean in its mathematical form through this more general way of comparing its values proportionally. And it brought its significance full circle by reintroducing the proportionality necessary to satisfy not only mental processes, but the very human impulse to discern aesthetics and beauty.

Golden Mean defines aesthetics and beauty.

2.3 A Word About Spirals

Our Geometers discovered as well that by using the same proportions, curvilinear constructs known as spirals could be created. They could establish a grid in which its square subdivisions were separated proportionately according to preordained relationships, such is 1:2 or 2:3, and then multiplied by other relationships such as $\sqrt{2}$ or $\sqrt{5}$ to create a curve of constantly changing proportions. While extremely sophisticated for the time, the spiral configurations were limited to the unreal world of flat space, or the plane. There was great significance to their creation, however, in that they were methods of using ninety-degree orientation and rational whole numbers to construct the irrationality of compound curvature.

But our friends were on the trail of something even grander. They figured out, but were unable to fully develop, the manner in which the sameness of the units within their patterns were linked with the asymmetry of the physical world, i.e., the senses. It was late in the 20th century, however, before this

THE FIBONACCI SERIES AND THE GOLDEN MEAN

The Fibonacci series is a sequence of numbers in which each succeeding number is equal to the sum of the last number and the number preceding it:

0, 1, 1, 2, 3, 5, 8, 13, 21, 34, 55, 89, 144, 233, 377, 610, 987......,etc.

If each succeeding number is compared with its prior number, we see that the relationships quickly converge on a unique ratio equal to:

$$\phi = \frac{\sqrt{5}+1}{2} = 1.618, \quad \text{or} \quad 1/\phi = 0.618$$

3/5	= 0.6	5/3	= 1.667
5/8	= 0.625	8/5	= 1.600
8/13	= 0.615	13/8	= 1.625
13/21	= 0.619	21/13	= 1.615
21/34	= 0.618	34/21	= 1.619
34/55	= 0.618	55/34	= 1.618
55/89	= 0.618	89/55	= 1.618
89/144	= 0.618	144/89	= 1.618
144/233	= 0.618 etc.	233/144	= 1.618, etc.

For the Golden Mean, in which the value of ϕ is powered sequentially as it distances itself from Unity, or 1, the generalized version of its sequence is:

$$........1/\phi^4, 1/\phi^3, 1/\phi^2, 1/\phi, 1, \phi, \phi^2, \phi^3, \phi^4........\text{etc.},$$

when numbers are substituted in the proportions, they equal:

```
   1.618-→    1.618-→    1.618-→    1.618-→    1.618-→    1.618-→
 .146       .236       .382       .618         1         1.618
  ←-0.618    ←-0.618    ←-0.618    ←-0.618     ←-0.618    ←-0.618

              1.618-→              1.618-→
 2.618       4.235                6.854
  ←-0.618                ←-0.618
```

The numbers above and below the numerical equivalent (in bold) of the sequence indicate that, ascending proportionally and descending proportionally (the inverse of ascending proportionally), these values differ by values of ϕ and $1/\phi$. This hints that its powered values increase additively.

Figure 17

See Figure 18

Linking spirals to Chaos Theory

Synthesis - a thought for the future.

linking was solidified with the aid of computer-generated calculations through the evolving science of Chaos Theory.

And a much more significant jump was made through this science by bringing Chaos' never-recurring-the-same-way generations into the deeper world of three-dimensional space. By so doing they moved to complete the task begun two millennia earlier of portraying, not only conceptual patterns, but their interplay within the energetic world in a rationally-based manner. **Not verbalized then and hardly mentioned today, is the additional byproduct of moving one more step into the process of synthesizing the worlds of the rational and of the intuitive, of left and right brain.**

Figure 18

From a process of sequencing configurations of increasingly large squares about each other, curves of various centers could be connected in a manner which created the spiral. Its outward growth from an "inner" core was taken to represent a model for growth, but an energetic model. As shown it could be described, in a ninety-degree orientation, by a series of whole-number relationships.

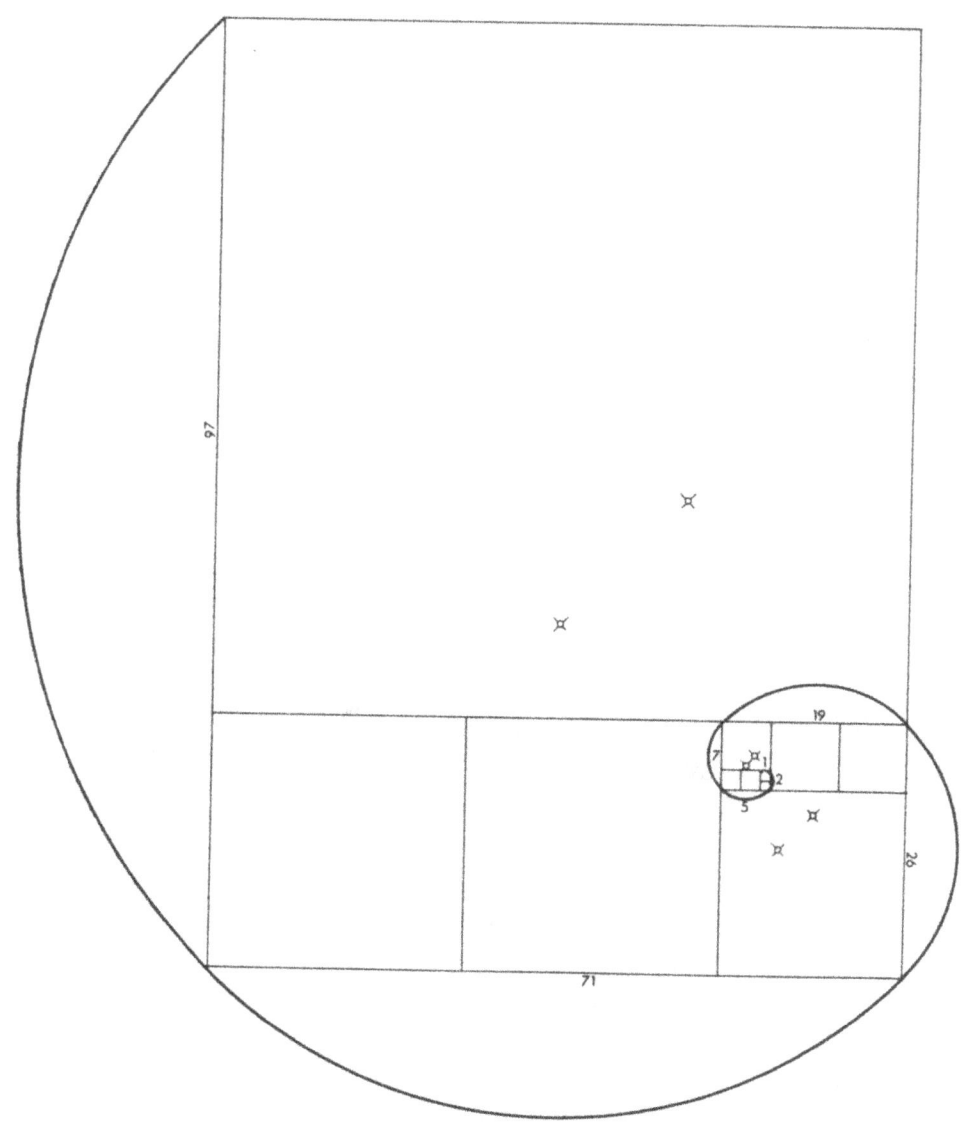

THE GEOMETRY OF UNIVERSAL MIND

2.4 A Thought For Future Consideration: The Mystery of Thoth

Did Geometry's evolving abstractions represent reality?

All of the above concepts, as the Piscean Age developed, grew increasingly esoteric and abstract. There were answers, to be sure, for all the questions the Geometric constructs and their philosophical implications generated. But were these abstractions real? Could a subdivision of matter really retain an innate sense of its original Unity? Could there really be some form of inner multiplication not apparent on the surface? Were conceptual patterns truly linked with the seemingly chaotic, free-will nature of the created Universe?

Science in the ensuing centuries indeed came to grapple with these concerns as demonstrable phenomena in nature. In the volumes to follow we will see how science has extended our sense of the nature of what is real, revealing the ancient Geometers' speculations to be surprisingly accurate. But this process was to have its own surprises as it unfolded, as Universal Mind slowly cycled earth's slant on reality back toward the holistic, intuitive characteristics of right brain predomination.

Among the mysteries inherited over the ages was a rather arcane statement attributed to the ancient god, Thoth, of Egyptian mythology. Obviously applicable to the abstractions with which our Geometer friends grappled, particularly those related to number, Thoth was to have said:

Thoth's paradox

I am One which transforms to Two
I am Two which transforms to Four
I am Four which transforms to Eight...
After all this, I am One.

To thinkers pondering this paradox, the numerical sequence from 1 to 2 to 4 to 8 described many things: Multiplication as subdivided Unity, as well as the implicit process of Transformation. Two represented complementarity and polarity. Two moving to four was the squaring process, generating area, possibly an implicit symbol of conceptual space. Four to eight extended area: eight was two^3, representing volume, possibly incorporating the parallel grid of time. But "After all this I am One?" Aside from some vague idea of a retained sense of Unity within the processes of Nature, what could this final line indicate? And how could it possibly be demonstrated? In the volumes to follow, we hope to find out.

2.5 Exercises

In order to empathize as deeply as possible with the manner in which Universal Mind manifested its Geometry, particularly over the past 2000 years, try the following exercises:

• Re-create the yantra of Figure 11. Its significance may deepen as you do so.

• What do terms such as Multiplication by Propagation and by Minification now mean to you? Can you represent them visually? Poetically? Geometrically? Try to portray them in a variety of such forms.

• As if you were doodling on a scrap of paper during a telephone conversation, create a Geometric shape that seems to have particular significance to you. Don't predetermine a shape; just doodle until you have something created that resonates with you. It can be simple or complex, made of straight or curved lines, or both. Then describe in a sentence or phrase the manner in which this shape is significant to you. That is, how does it describe you?

• If you have some acquaintance with traditional philosophy, what do the various Geometric constructs of this book now mean to you in philosophic terms? Do you sense meanings other than those you're familiar with? Set them down in a notebook for future reference.

Appendices

APPENDIX "A"
THE TWELVE DEGREES OF FREEDOM

Degree	Function	Process Description	Focus
1 / Power	Creation	Purpose, Identity Individuality	Inner
2 / Structure	Order	Experience, Wisdom Expression of Love	Outer
3 / Reflection	Inner Abstraction	Trust, Thought, Language Present/Past	Inner
4 / Adaptability	Incentive	Stimulation to Activity Direction, Preference	Outer
5 / Desire	Feelings separated as interest and indifference	Choice, Emphasis, Feeling preference	Inner
6 / Repulsion/attraction	Discrimination	Realization of individuality Identity = Activity	Outer
7 / Concreteness	Focus	Vitality, Clear desire Identity = Knowledge	Inner
8 / Abstractness	Conceptuality (Geometry)	Synthesis, Quantum Conceptual knowing	Outer
9 / Individual life	Individuation	Power in Relationship Identity = Relationship	Inner
10 / Divine Life	Composite Identity	Masc/Fem Synthesis Teach, Extend self	Outer
11 / Divine self/otherness	Wholeness	Prime Complementarity	*******
12 / One-ness	Purity	No Dualities	*******
**** Foreverness	***Undefined***	***Undefined***	*******

Note: For any system coming into existence, whether inanimate or human, there is a progression within which individuation is born, matures, and finds its fulfillment in its surrender to the wholeness of Mind. The above is a representation based on traits familiar to human experience, which expresses both the flower-ing of individuality and its re-unification (with identity intact) into the undefined nature of the whole. After this, there is nothing relative, only absolute quality of pure life.

Notes

APPENDIX "B" CHARACTERISTICS OF ASTROLOGICAL PERIODS

SIGN	PREDOMINATION	MIND RELATED CHARACTERISTICS
Capricorn	Left Brain	Utility, Nimbleness, Imagination
Aquarius	Right Brain	Rest/rebirth, Fertility
Pisces	Left Brain	Power, Focus, Secrecy, Desire/manifestation
Aries	Right Brain	Giving nature, Cunning, Beauty
Taurus	Left Brain	Strength, Work, Austerity
Gemini	Right Brain	Conflict, Polarity, Adventure
Cancer	Left Brain	Ritual, Inertia, Sacredness
Leo	Right Brain	Anger, Ferocity/protectiveness
Virgo	Left Brain	Justice, Sternness, Protection
Libra	Right Brain	Balance/complementarity, Bargaining
Scorpio	Left Brain	Separation, Dark/light, Attachment
Sagittarius	Right Brain	Wisdom, Warring, Healing

NOTE: The above were compiled from the myths of the signs as traits of each aspect of the cycles of Earth's precessional orbiting. Comparisons can also be made to the months of the year and their traits, as well as the different seasons of life.

Notes

APPENDIX C

Cultures		1		2	7 9 3 4 5 6	8	10	11	12
Ages	Cancer	Gemini		Taurus	Aries		Pisces		Aquarius

1. Early Egyptian – First hieroglyphic writing, first monuments

2. Egypt, Old Kingdom – Commerce, mining. First use of stone in lieu of brick. Engineering, architecture, sculpture, painting, navigation, industrial arts and sciences, astronomy. Strong centralized government. Evidence of strong organizational, skills, effective bureaucracy.

3. Egypt, Middle Kingdom – Religion is democratized. King as "good shepherd." Renaissance of culture. Golden age of Egyptian literature. Art, jewelry, architecture.

4. Sumerian – Pantheon of gods. Literature – mythology, epic poetry. Cultural prominence of love, sex, war. Agriculture is mainstay of economy.

5. Babylonian – Agricultural society. Use of metal in artisanship. Use of dyes, bleaches, paints, perfumes. Cuneiform writing basis of administrative, literary institutions. Strong, warlike cultural traits alternating with political chaos and disintegration.

6. Assyrian – Begins with strife due to Sea People from Western Asia. War. Beginnings of scientific, economic administrative policies from the Code of Hammurabi. Literature – Myths, epics. Poetry prominent in historical chronicles, romances, social issues. A deep awareness of and interaction with nature.

7. Egypt, New Kingdom. Elevated roles for women. Art and architecture prevail. Constant war, uprisings. Failure of monotheistic religion. The beginning of Egypt's domination by Rome and Byzantium.

8. Celtic – Tribal Agricultural society. No political unity. Feasting drinking, fighting.

9. Greek – Warlike, tribal. Mythological poetic literature. Beginnings of rational thought as basis of law, science, philosophy.

10. Roman – Strong, centralized society. Law architecture, religion, philosophy influenced by Greek culture.

Notes

APPENDIX C (Cont'd)

11. European Renaissance – Vital economic Structure, travel, commerce. Rational-based religion. Vital artistic, literary culture. New developments in architecture. Beginning of democratized social institutions.

12. Twentieth Century industrial society – Movement from agricultural to urban, industrial society. Democratization of technology. General material wealth. Existential philosophy. The novel and short story as literary forms. Great social democratization. Emergence of the individual in society. Technology in art forms – movies, music.

Notes

Appendix "D"

				Edges	Faces	Vertices	Length
Tetrahedron				6	4	4	$\sqrt{2}$
Octahedron				12	8	6	$\frac{1}{\sqrt{2}}$
Cube				12	6	8	1
Icosahedron				30	20	12	ϕ
Dodecahedron				30	12	20	$\frac{1}{\phi}$

Notes

APPENDIX "E"

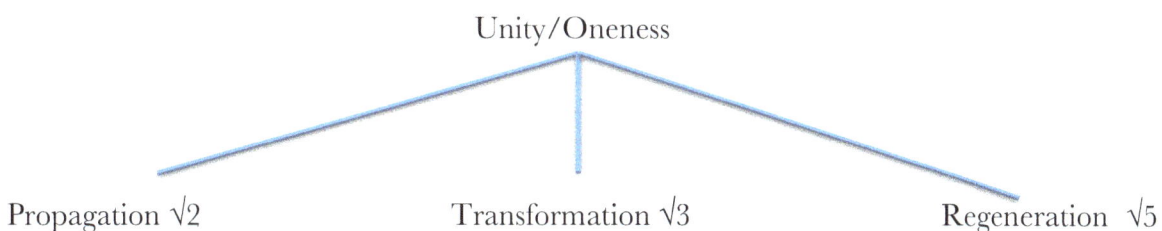

Historical Symbolism for Natural Processes

In this, the traditionally accepted manner of representing natural processes with Mathematical and Geometrical Symbology, we are answering in both a rational and a non-linear way this question: How is a diverse Universe created from the notion of One-ness? The interplay of natural forces has been traditionally reduced to three concepts:

Propagation – Represented by the square root of three, or the diagonal of a square. Implies area. Within this proportionality lies the accepted basis of Nature's ability to subdivide or re-create Oneness in a linear way. This involves making many-of-One as well as numerical increase, i.e., creating Two from One, Four from Two, etc.

Transformation – Represented by the square root of three, the diagonal of the cube. Implies volume in lieu of area. Proportional constant is also found in the *Vesica Piscis*, which points the way toward an underlying process by which Propagation occurs. This is the recognition that there is an internal restructuring, whether it be atomic, Geometric, or others, by which Universe creates and perpetuates diversity.

Regeneration – Represented by the square root of five, or the diagonal of two squares. This is a concept not well portrayed by conventional Geometry, but which recognizes the appearance- disappearance- reappearance phenomenon in Nature. Colored largely by focused exclusionary left-brain viewing of natural processes, it tacitly creates death or disappearance, and its double, isolation, without affirming the process of continuity by which Nature works in a most efficient manner.

Notes

Appendix "E" Cont'd

Complementarities lie inside of other complementarities…

…and function energetically through processes involving scale, spiraled inwardly by recursion or energetic spin.

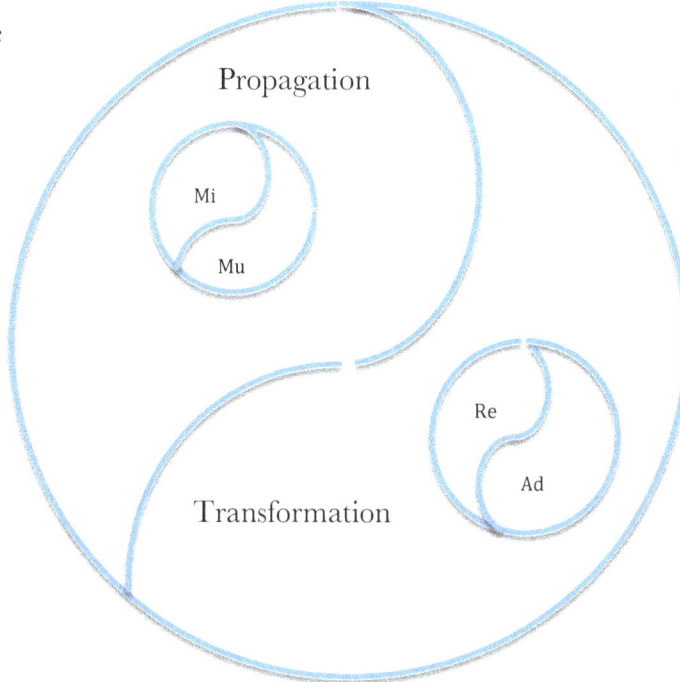

The new model at the basis of natural processes is seated in the concept of complementarity. What this implies is that the inherent duality, the field of opposites within Nature, is not a division of Unity/Oneness that is set against itself. It is through this Twoness, this ultimate equality, that Nature is able to function. From this Twoness comes conceptual Threeness, the half-quantum, which is at the basis of Nature's Geometric structure. We can re-configure our ancient Geometers' identification of these natural processes thusly:

Propagation – the outward nature of the creation of holographic diversity from Unity/Oneness. It occurs through the sixty-degree orientation Geometry of The Matrix on any dimensional level and, by its symmetry, connects all dimensional levels.
    Multiplication – the numerical increase in diversity's expression. It can be additive or multiplicative.
    Minification – the ability to use scale to further create diversity "within" and "without" the Geometry of any one dimensional level.
Transformation – The inward nature of the creation of holographic diversity. By the efficient Geometric structuring of The Matrix it gives conceptual space and its complement, space-time, its fluidity by using The Matrix's Geometry to allow the focus of consciousness to enfold and unfold its nature implicately and explicately.
    Regeneration – the alteration of phenomenal effects across dimensional boundaries.
    Adaptation – the ability to change Geometric shapes as energy is added to and deleted from any phenomenal effect.

Notes

CREDITS

The following publications have inspired this work and those that follow-many illustrations and some text have been adapted therefrom. Please read ahead. Enjoy!

Itzhak Bentov, Stalking the Wild Pendulum, Bantam, N.Y. 1981
J.F. Bierlein, Parallel Myths, Ballantine Books, N.Y. 1994
Fritjof Capra/David Steindl-Rast, Belonging to the Universe, Harper, San Francisco, 1991
Fritjof Capra, The Tao of Physics, Shambhala Publ., Boulder, CO, 1976
Keith Critchlow, Order in Space, Viking Press, N.Y.
H.M. Cundy/ A.P. Rollett, Mathematical Models, Oxford Univ. Press, London, 1972 Albert Einstein, Relativity, Crown Publ. N.Y.
Buckminster Fuller, Synergetics I&II, Macmillan Publishing Co., Inc., N.Y. 1975, 1979
James Gleick, Chaos, Penguin Books, N.Y. 1987
Alice O. Howell, The Heavens Declare, pp. 145-146
Robert Lawlor, Sacred Geometry, Thames and Hudson, Ltd., London, 1982 Henrietta McCall, Mesopotamian Myths, Univ. of Texas Press, Austin, 1990
John Mitchell, The Earth Spirit, Thames And Hudson, N.Y. 1975
Peter Mollman, Wonders of the Universe, World Book, Inc. Chicago, 1989
P.D. Ouspensky, A New Model of the Universe, Vintage Books, N.Y. 1971
H.A. Rey, The Stars, Houghton Mifflin, Boston, 1976
Bertrand Russell, Wisdom of The West, Crescent Books, London, 1977
Ruth Shephard/Barry Sanders, The Sacred Paw, The Penguin Group, N.Y. 1985 Julius D.W. Staal, The New Patterns in the Sky, MacDonald & Woodward Publ., Blacksburg, Va. 1988
Michael Talbot, Beyond the Quantum, Bantam Books, N.Y. 1987
Michael Talbot, The Holographic Universe, Harper Perennial
Jean-Pierre Verdet, The Sky, Mystery, Magic and Myth, Harry M. Abrams, Inc. N.Y.
Margaret J. Wheatley, Leadership and the New Science, Berrett-Koehler, San Francisco, 1994
David Gordon White, Myths of the Dog-Man, Univ. of Chicago Press, Chicago, 1991 Ken Wilber, The Spectrum of Consciousness, Theosophical Publishing, London, 1977 Eva Wilson, North American Indian Designs, Dover Publications, N.Y. 1984
Daniel Winter & Friends, Sacred Geometry, The Alphabet of the Heart, Crystal Hill Farm, Eden, N.Y.
Fred Allen Wolf, Parallel Universes, Touchstone, N.Y. 1988
Arthur M. Young, The Geometry of Meaning, Robert Briggs Associates, 1976
Sri Yukteswar, The Holy Science, Self-realization Fellowship, pp. ix-xix

Notes

Index

Notes

ALPHABETIC INDEX

                                                                    Page

Adaptation	Appendix "E"
Alternation	29
Anasazi	16
Asymmetry	6
Attractors	5, Fig 1, 15
Aquarius	8, Appendix "B", Appendix "C"
Babylonian	Appendix "C"
Beauty	5, 26
Big picture	2, 13
Cancer (Sign)	Fig 2, Appendix, "B", Appendix "C"
Capricorn	Appendix "B", Appendix "C"
Celtic	Fig 8, Appendix "C"
Equinox	Fig. 2
Fibonacci Series	27
Forever	Appendix "A"
FOUR, Fourness	17, 19, Appendix "E"
Four D (4D)	Fig. 1
Fragmentation	10
Gemini	Fig. 2, Appendix "B", Appendix "C"
Geometry	9, 14, 15, 16, 17, 20, Fig. 10, Appendix "A", Appendix "E"
Gods	9
Golden Mean	24, Fig. 16, 26, Fig. 17
Greek	9, Fig. 9, 19, Fig. 13, 22, 25, 36, Appendix "B"
Grid	Fig. 1
Holism	11

Notes

Index, Cont'd

	Page
Holography	Fig 1
Icosahedron	Appendix "D"
Integrated observer	11
Intellect	8
Intuition	8, 11
Japanese calligraphy	Fig 11
Left brain	10, 12, 14, Appendix "B"
Leo	Fig 2, Appendix "B"
Libra	Fig 2, Appendix "B"
Lile	Appendix "B"
Love	7, Appendix "A"
Macrocosm	8
Mathematics	14, 17, 23, Fig 15, 26, 27
Meaning	1, 2, 3
Metaphysical	5, 6, 14
Mind	4, 9
Minification	Appendix "E"
Models	Appendix "E"
Multiplication	20, 21, 22, 23, 24, Fig. 12, Fig. 15, Fig. 16, 29, Appendix "E"
Mythology	16, 17, Appendix "C"
Native American	Fig. 10
New Kingdom	16, Appendix "C"
Ninety degree orientation	19, Fig. 18
Octagon	Fig. 9
Octahedron	Fig. 8, 17

Notes

Index, Cont'd

        Page

Old Kingdom .... 16, Appendix "C"

ONE, oneness .... 7, Fig 16, 29, Appendix "A", Appendix "E"

Passion 12

Passive observer 10

Patterns 5

Pentagon .... 23, Fig 14

Perception .... 33, Fig 15

Perspective 25

Pi 26

Pisces, Piscean Age .... Fig. 2, 9, 17, 24, 29, Appendix "B", Appendix "C"

Plato 17

Powers .... Fig. 17, Appendix "A", Appendix "B"

Precession 9

Propagation .... Fig 16, Appendix "E"

Proportion .... 25, Fig. 17, 27

Pyramid .... Fig. 8, 18, 19

Pythagoras .... 18, Fig. 13, Fig. 14

Quantum, quanta 24

Ratio 25

Regeneration .... 262, Appendix "E"

Relationships .... 2, 23, Fig. 15, Fig. 18, Appendix "A"

Renaissance .... Appendix "B"

Right brain .... 8, 16, 17, 18, 29 Appendix "B"

Sagittarius .... Fig. 2, Appendix "B", Appendix "C"

Scorpio .... Fig. 2, Appendix "B", Appendix "C"

Notes

Index, Cont'd

	Page
Sex	12
Slowness	10
Solstices	Fig. 2
Spiral	16, Fig. 8, 28. Fig. 18
Square root of five	23, Fig. 14, 28
Square root of three	21, Appendix "E"
Square root of two	Fig. 12, 19, 25, 26, 28
Sumerian	16, 17, Appendix "C"
Sun	Fig. 2
Symmetry	Fig. 16
Taurus	Fig. 2, Appendix "B", Appendix "C"
Technology	Appendix "C"
Thales	9, 17, 25, Fig. 16
Thoth	29
THREE, threeness	17, 21
Transformation	20, 22, Appendix "E"
TWO, Twoness	17, 18, Appendix "E"
TwoD, (2D)	Fig. 1
Unity	17, 19, 20, 23, Fig. 15, Fig. 16, 27
Universal Mind	24, 29
Universe	27 Appendix "E"
Vesica Piscis	23 29, Appendix "E"
Virgo	Fig. 2, Appendix "B", Appendix "C"
"Window"	26

Notes

Index, Cont'd

                                                                                                                       Page

Yantra .... 16, Fig 11, 30

Zeno 18

Ziggurat .... 16, Fig. 10

Zodiac .... Fig 2, 9

Notes

ABOUT THE AUTHOR

Bob Mustin has had a brief naval career and a longer one as a civil engineer and has been a North Carolina Writers Network writer-in-residence at Peace College under the late Doris Betts' guiding hand. In the early 90s he was the editor of a small literary journal, The Rural Sophisticate, based in Georgia. His work has appeared extensively in print and electronic publications

To learn more about Bob Mustin, visit:
Website: www.bobmustin.com
Blog: bobmust.wordpress.com

www.ingramcontent.com/pod-product-compliance
Lightning Source LLC
Chambersburg PA
CBHW061150070526
44584CB00034B/4476